Plant Proteins for Human Food

NUTRITION SCIENCES

Also in this series:

Gammaster, Food Irradiation Now. Proceedings of a Symposium held in Ede, The Netherlands. ISBN 90-247-2703-0

In preparation:

Jansen LPBM and Zuilichem DJ, Food Extrusion Technology

Plant Proteins for Human Food

Proceedings of a European Congress
held in Nantes, France, 5–7 October 1981

Organized by
Institut National de la Recherche Agronomique

Edited by

C. E. Bodwell
Chief, Protein Nutrition Laboratory
Beltsville Human Nutrition Research Center
USDA, Beltsville, MD
U.S.A.

L. Petit
Institut National de la Recherche Agronomique
Laboratoire de Biochemie des Aliments
Nantes-Cedex, France

Coordinated by
J. Gueguen

Institut National de la Recherche Agronomique
Laboratoire de Biochemie des Aliments
Nantes-Cedex, France

Reprinted from *Qualitas Plantarum: Plant Foods For Human Nutrition,* Vol. 32, Nos. 3/4

1983 **MARTINUS NIJHOFF/DR W. JUNK PUBLISHERS**
a member of the KLUWER ACADEMIC PUBLISHERS GROUP
THE HAGUE / BOSTON / LANCASTER

Distributors

for the United States and Canada: Kluwer Boston, Inc., 190 Old Derby Street, Hingham, MA 02043, USA
for all other countries: Kluwer Academic Publishers Group, Distribution Center, P.O.Box 322, 3300 AH Dordrecht, The Netherlands

Library of Congress Cataloging in Publication Data

Main entry under title:

Plant proteins for human food.

(Nutrition sciences ; v. 2)
Includes bibliographical references and index.
1. Plant proteins--Congresses. I. Bodwell, C. E.
II. Petit, L. III. Guéguen, J. IV. Institut national
de la recherche agronomique (France) V. Series.
[DNLM: 1. Plant proteins--Congresses. 2. Nutritive
value--Congresses. 3. Food supply--Congresses. QU 55
P713 1981]
TP453.P7P55 1983 664'.64 83-11424
ISBN 90-247-2856-8

ISBN 90-247-2856-8 (this volume)
ISBN 90-247-2716-2 (series)

PRINTED IN THE NETHERLANDS

Contents

PART III. EXTRACTION PROCESSES AND END-PRODUCT CHARACTERISTICS

PART IV. FUNCTIONAL PROPERTIES

Preface

C.E. Bodwell and L. Petite

Until a few years ago, a lack of dietary protein was considered to be the primary cause of diet-related malnutrition in the world. This view has changed and most would now agree that if calorie needs are fulfilled, then protein intakes will be adequate. Although true, this statement does not negate the importance of providing sufficient protein in the diet. The development of an adequate world supply of dietary protein requires an increased utilization of plant proteins directly by humans.

Adequate supplies of dietary protein do exist if all of the plant protein which is produced each year and is edible for humans, were actually consumed in an equitable manner by the human population. However, from a practical viewpoint, this is unlikely to occur. A large proportion of the available edible plant protein will undoubtedly continue to be fed to livestock because of the demand for animal protein for human consumption. For major shifts to occur in the utilization of plant protein, suitable vegetable protein products must be developed. Some are (or will be) similar to, and compete with, animal protein products as desirable components of the diet. In addition, however, it is very important that new products be developed which will be accepted as preferred food items on their own merit and not merely exist as imitations or extenders of animal protein products. The subject of the Congress on which this book is based is thus timely and of major importance. This book consists of five parts.

In Part I, 'The Importance of Plant Proteins', the problem of the increased consumption by humans of animal protein which occurs concurrent with an increase in standard of living is documented. A basis is clearly established for the need for developing improved plant protein products for direct consumption by humans.

In Part II, 'Extraction Processes and End-Product Characteristics', procedures used for extracting and concentrating protein from legumes, oilseeds and green leaves are reviewed. Nutritional and other characteristics are generally considered.

In Part IV, 'Functional Properties', fundamental concepts, as well as practical considerations, of the functionality of various plant protein materials are discussed. Included is a consideration of improving functional properties by enzymic modification.

In Part V, 'Nutritional Aspects', the influence of processing on the nutritional value of plant proteins is briefly reviewed and the antinutritional factors which are present in most plant proteins are discussed.

This book is broad in scope, yet it provides sufficient detailed information to make it useful to a wide range of readers, including students, food scientists, nutritionists, agronomists, and generally to anyone who seeks information on plant protein sources, processing, or utilization.

Contributors

J. Adler-Nissen, NOVA INDUSTRI A/S, Enzyme Applications R & D, Bagsvaerd, Denmark
J. Baudet, Laboratoire d'Etude des Proteines, Department de Physiologie et Biochimie Vegetales, I.N.R.A., Versailles, France
D. Boulter, Department of Botany, University of Durham, Durham DH1 3LE, England
R. Costantino, Ministère de l'Agriculture, Institut National de la Recherche Agronomique, Centre Agro-Alimentaire, Laboratoire de Biochimie des Aliments, Cedex, France
S. Eriksen, NOVO INDUSTRI A/S, Enzyme Applications R & D, Bagsvaerd, Denmark
P. Fantozzi, Instituto di Tecnologie Alimentari, Universita degli Studi di Udine, Udine, Italy
G. Fauconneau, Institut National de la Recherche Agronomique, Paris, France
R. Ferrando, 107 rue de Reuilly, Paris, France
P.A. Finot, Nestlé Products Technical Assistance Co. Ltd., Research Department, La Tour-de-Peilz, Switzerland
R. Fiorentini, Instituto di Industrie Agrarie dell'Universita, Via S. Michele degli Scalzi, Pisa, Italy
C. Galoppini, Instituto di Industrie Agrarie dell'Universita, Via S. Michele degli Scalzi, Pisa, Italy
C. Giddey, Battelle, Geneva Research Centers, Carouge-Geneva, Switzerland
S. Grillo, Instituto di Agronomia Generale e Coltivazione Erbacee, Catterdra di Niglioramento Genatico delle Piante Coltivate, Universita degli Studi di Hapoli, Portici, Italy
J.Q. Gueguen, Ministère de l'Agriculture, Institut National de la Recherche Agronimique, Centre Agro-Alimentaire, Laboratoire de Biochimie des Aliments, Cedex, France
A.-M. Hermansson, SIK, The Swedish Food Institute, Göteborg, Sweden
P.J. Lillford, Unilever Central Resources, Colworth House, Sharnbrook, Bedford, England
L.M. Monti, Instituto di Agronomia Generale e Coltivazione Erbacee, Catterdra di Miglioramento delle Piante Coltivate, Universita degli Studi di Napoli, Portici, Italy
J. Mossé, Laboratoire d'Etude des Proteines, Department de Physiologie et Biochimie Vegetales, I.N.R.A., Versailles, France
H.S. Olsen, NOVO INDUSTRI A/S, Enzyme Applications R & D, Bagsvaerd, Denmark
A. Sensidoni, Instituto di Tecnologie Alimentari, Universita degli Studi di Udine, Udine, Italy
G. Sodini, ASSORENI, ENI Group Research Association, 00015 Monterotondo, Rome, Italy
L. Tranchino, ASSORENI, ENI Group Research Association, 00015 Monterotondo, Rome, Italy
D.J. Wright, A.R.C. Food Research Institute, Division of Chemistry and Biochemistry, Norwich, England

Qual Plant Plant Foods Hum Nutr 32 (1983) 205–223

World protein supplies: the role of plant protein

GUY FAUCONNEAU

Institut National de la Recherche Agronomique, 149 rue de Grenelle, 75341 Paris Cedex 07, France

Key words: plant protein, animal protein, world protein supplies, future trends

Abstract. The per capita levels of protein from various animal sources and from plant sources which are currently available for human consumption are presented and discussed. Changes in the patterns of consumption which occur as the standard of living is improved are documented for various countries. These patterns are used to suggest that increased production of plant protein alone will not necessarily be a solution to the problem of meeting increasing world needs for dietary protein. Products from plant protein sources need to be developed which will compete with animal protein sources.

Introduction

The amounts of protein available for consumption (but not necessarily actually consumed) in the various regions of the world vary from 50 to 100 g per day per inhabitant for populations living under very different socio-economic conditions [2–10]. For half of the world population, only about 50 g of protein per year per inhabitant is available, of which 80% is plant protein. On the other hand, for one fourth of humanity, the average amount of protein available is 90 to 100 g per day per inhabitant of which more than half is of animal origin. In relation to the total quantity of available energy, protein represents from 10 to 15% of available calories, but this value may reach 20% if estimated relative to the energy actually consumed in the developed countries.

Present consumption of proteins of different origin

In the industrialized OECD countries and Eastern Europe (1,150,000,000 inhabitants, USSR included), the populations have, on an average, 90 to 100 g of protein per day per inhabitant at their disposal, of which more than half is of animal origin (Table 1). Some of the richest countries (USA, New Zealand, Australia, France) consume almost 3/4 of their protein in the form of animal protein. These countries have more than 3,000 Kcal per day per inhabitant to satisfy the energy requirements of their populations.

Table 1. Protein consumption (g/day per capita)

Population (10^{-6})			Total	Plant protein	Animal protein
	World				
3290		1961–1965	66	44.1	21.8
4043[a]		1974	69	44.6	24.4
4250[b]		1977	69.4	44.8	24.6
	Developed countries				
690	Total	1961–1965	90	41.3	48.8
762[a]		1974	95.5	39.2	56.2
775[b]		1977	96.4	39.4	57.0
	USA				
194		1961–1965	101.8	33	68.8
210[a]		1974	104.3	32.6	71.7
218[b]		1977	106.2	33.5	72.7
	France				
48.7		1961–1965	96.9	44.2	52.6
52.5[a]		1974	100.0	35.9	64.1
53.3[b]		1977	103.0	37.0	66.0
	Centrally planned countries				
	Europe				
333		1961–1965	94.5	56.5	38.0
366[a]		1974	100.5	52.2	48.3
372[b]		1977	103.0	51.6	51.4
	USSR				
230		1961–1965	96.2	58.0	38.0
256[a]		1974	102.0	52.5	49.5
261[b]		1977	103.5	52.0	51.5
	Asia				
764		1961–1965	55.2	44.0	11.3
923[a]		1974	61.0	48.0	13.0
1000[b]		1977	63.1	49.7	13.4
	Developing countries				
1500	Total	1961–1965	53.5	42.7	10.8
1991[a]		1974	54.1	42.5	11.6
2100[b]		1977	55.4	43.5	11.9
	Africa				
246		1961–1965	53.5	42.7	10.8
328[a]		1974	53.7	43.2	10.5
350[b]		1977	54.9	44.3	10.6
	Latin America				
247		1961–1965	64.1	39.3	24.8
333[a]		1974	64.5	39.3	25.2
349[b]		1977	65.2	38.5	26.7
	Asia				
854		1961–1965	49.1	42.5	6.6
1105[a]		1974	49.1	41.6	7.5
1186[b]		1977	49.6	42.0	7.6

Source: FAO [1] [a] 1975 Estimate [b] 1978 Estimate.

In the poorest developing countries (1,550,000,000 inhabitants: South-eastern Asia, Africa), the populations have only 50 g protein per day per inhabitant at their disposal of which less than 20% is of animal origin (Asia: 15%; Africa: 19%).

Table 2. Protein and fat consumption (g/day per capita)[a]

Country	Total proteins (animal + plant)	Plant protein	Animal protein	Fats
Europe				
1955–59	90.0	52.0	38.0	
1969–70	92.5	44.4	48.1	126.0
1976–77	96.0	43.2	52.8	134.0
Denmark				
1955–59	84.0	32.7	51.3	153.0
1969–71	88.0	32 0	56.0	162.0
1976–77	89.9	30.9	59.0	162.0
Western Germany				
1955–59	83.1	38.1	45.0	143.0
1969–70	86.0	32.0	54.0	153.0
1976–77	88.0	30.0	58.0	155.0
Great Britain				
1955–59	89.8	37.7	52.1	132.0
1969–70	93.9	36.7	57.2	148.0
1976–77	91.7	35.7	56.0	139.0
Italy				
1955–59	81.2	53.2	28.0	
1969–71	96.0	56.0	40.0	120.0
1976–78	98.1	53.4	44.7	123.0
Sweden				
1955–59	85.9	30.8	55.1	130.0
1969–70	84.0	26.0	58.0	132.0
1976–77	92.4	30.6	61.8	139.7
Switzerland				
1955–59	92.3	41.5	50.8	127.0
1969–70	90.0	35.0	55.0	150.0
1976–77	88.0	31.7	56.3	154.7
Spain				
1955–59	74.4	52.2	22.2	73.4
1969–70	82.0	43.0	40.0	100.0
1976–77	93.5	44.9	48.6	127.0
Greece				
1955–59	97.6	69.2	28.4	86.0
1969–70	101.3	56.6	44.7	109.0
1976–77	104.2	55.5	48.7	133.0
Japan				
1955–59	68.3	53.2	15 1	26.0
1969–70	79.0	48.0	31.0	59.0
1976–78	86.4	44.6	41.8	73.0
New-Zealand				
1955–59	102.3	34.2	68.1	144.0
1969–70	108.2	34.0	74.2	148.0
1976–77	110.0	33.2	76.8	149.5

[a]Source: FAO [2], OECD [4].

Between these two extremes, efforts are being made in Asia (and, in particular, in China with a little more than one billion inhabitants) to increase the availability of protein which at present is more than 60 g of protein per inhabitant per day of which more than 20% is in the form of animal protein.

In the industrialized countries, the food habits have changed during the last thirty years with changes in life styles. The fast urbanization and the development of secondary and tertiary sectors in the industrialized societies have made a great part of the populations sedentary; the decrease in physical activities has been accentuated by the development of motorized transport and escalators.

According to the development of the various sources of available protein in the different industrialized countries (Tables 1, 2), the changes of food habits towards an increasing consumption of animal protein have reached different steps in different countries. The industrialized countries have first satisfied their main requirements, especially for energy, up to 3000 Kcal per year per inhabitant; thereafter, as their standard of living increased, the different industrialized countries of OECD and Eastern Europe increased the availability of total proteins gradually. This was followed by an increase in the availability of animal protein.

In the most developed countries, like the USA, the consumption of plant protein had already regressed before World War II and this trend has persisted since 1950. A similar development could be observed in most European countries since 1950: a decrease in the consumption of cereals (bread and noodles), starchy food (potatoes), and leguminous plants (beans, peas, etc.), and an increase in the consumption of products of animal origin. From 1960, the availability of protein per day per inhabitant in the USA and New Zealand has equaled 100 g, of which two thirds has been of animal origin. During the same period, most countries in Western Europe had only 85 to 90 g of protein at their disposal, half of which was of animal origin. However, the share of animal protein only represented 28 g out of 85 g of total protein in Italy, 22 g out of 74 g total in Spain, 28 g out of 97 g total in Greece and 15 g out of 68 g total in Japan. All of the industrialized countries primarily attempted to satisfy their quantitative requirements for total protein, reaching an average of about 90 g; then they increased their availability of animal proteins by a stepwise replacement of plant proteins by animal proteins, mainly composed of fish followed by eggs and dairy products. This trend has led to an increase in the availability of fats (Table 2), especially in Spain, Greece and Japan, since fats are often combined with animal proteins in meats such as pork and meat products as well as in dairy products (milk and cheese).

The distribution of the various sources of proteins in the industrialized countries (Table 3) also shows the different stages of this general consumption pattern. Some countries are still consuming 45–50 g of plant proteins (Italy, Spain, Portugal, Japan) either in the form of cereals (noodles and rice) in Italy and Portugal, or leguminous plants (pea, bean soybean) in Spain and Japan. The daily consumption of plant proteins in the most developed countries is much smaller: 25 g in Sweden, 28 g in Western Germany, 30 g in the USA and in Switzerland.

Eggs may supply up to 5 g of animal protein per day per capita. In the United States, this consumption was 5.5 g per day 20 years ago, but it has now been reduced to about 4.8 g per day.

Table 3. Daily consumption (1976–1978) of protein of various origin (g/day per capita)[a]

Source	Great Britain	France	Western Germany	Italy	Spain	Portugal	Japan	Sweden	Switzerland	USA
Meat	25.7	38.00	32.0	24.0	21.0	17.3	10.5	22.0	28.0	41.0
Eggs	4.3	3.75	5.2	3.4	5.0	1.3	4.2	3.8	3.4	4.8
Dairy products										
(cheese included)	23.2	24.00	20.0	18.5	14.5	8.5	5.1	36.0	26.7	25.0
Fish and shell fish	2.2	5.30	3.4	4.5	12.0	13.5	16.0	7.7	1.9	2.5
Vegetables and Fruits	9.6	12.00	8.5	12.6	18.0	15.3	20.0	7.5	8.9	11.7
Cereals	23.5	23.00	19.5	40.0	26.5	36.0	27.0	18.5	21.5	18.5
Total	88.5	105.05	90.6	104.0	97.0	92.4	82.8	95.5	91.4	103.5
Protein:										
animal origin	55.0	71.0	60.0	50.5	52.0	40.6	35.8	69.0	60.0	73.3
plant origin	33.1	34.0	30.0	53.5	44.5	51.3	47.0	26.5	31.0	30.2

[a]Source: OECD [4].

Table 4. Daily consumption (1976–1978) of fats of various origin (g/day per capita)[a]

Source	Great Britain	France	Western Germany	Italy	Spain	Portugal	Japan	Sweden	Switzerland	USA
Meat	46.0	50.0	70.0	35.0	33.5	27.7	19.0	47.0	60.0	62.0
Eggs	4.2	3.4	4.9	3.2	4.6	1.2	4.0	3.6	3.2	4.8
Dairy products (cheese included)	20.0	17.0	14.0	13.5	13.0	7.4	4.0	23.3	22.3	18.2
Fish and shell fish	0.9	2.0	1.3	1.6	4.0	4.0	4.6	2.8	0.9	1.0
Vegetables and fruits	2.4	5.0	3.2	5.5	7.5	5.5	5.4	2.2	4.7	6.6
Cereals	3.2	3.0	3.0	5.2	3.4	4.8	3.1	2.8	2.9	2.6
Fat and oils (butter included)	61.0	67.5	66.0	64.0	60.0	65.3	32.0	52.0	59.0	66.5
Total	140.0	148.0	166.0	128.0	126.0	117.0	72.5	133.7	155.0	162.0
Fats of:										
Animal origin	105.0	105.0	123.0	62.0	60.0	59.0	64.0	97.0	103.0	93.0
Plant origin	35.0	43.0	43.0	66.0	66.0	58.0	8.5	33.7	52.0	69.0

[a]Source: OECD [4].

Table 5. Daily consumption of energy (Kcal/day per capita)[a]

Source	Great Britain	France	Western Germany	Italy	Spain	Portugal	Japan	Sweden	Switzerland	USA
Meat	520	650	750	420	410	325	210	540	630	700
Eggs	55	50	68	45	65	18	55	50	45	62
Dairy products (cheese included)	380	305	280	240	240	135	85	520	400	370
Fish and shell fish	17	50	26	30	85	100	110	60	16	20
Vegetables and Fruits	320	375	380	360	580	400	355	310	340	340
Cereals	700	650	633	1250	810	1235	1200	600	650	600
Carbohydrates	520	400	400	340	300	300	270	450	410	475
Fat and oils (butter included)	540	650	600	560	540	585	285	450	520	585
Total[b]	3100	3150	3150	3250	3050	3120	2540	300	3100	3200
Calories of:										
Animal origin	1270	1400	1450	820	800	700	740	1350	1300	1300
Plant origin	1820	1750	1700	2430	2250	2420	1800	1650	1800	1900

[a]Source: OECD [4].
[b]Cacao products included, alcoholic beverages excluded.

In the industrialized countries of temperate climate, dairy products supply a large amount of animal protein: Sweden is traditionally the biggest consumer of dairy protein (36 g) followed by Switzerland (27 g) the USA, France and Great Britain. Some Mediterranean countries, traditionally small consumers of dairy products, are increasing their daily caput consumption: Italy has reached 18 g and Spain 14.5 g. Portugal still has only 8.5 g per day per inhabitant and Japan only 5 g per day per inhabitant.

Meat represents the largest increase among products rich in animal proteins. The overall amount of meat protein available per inhabitant per day represents 41 g in the USA, 38 g in France, 32 g in Western Germany, 28 g in Switzerland, 25 g in Great Britain, 24 g in Italy, 22 g in Sweden and 21 g in Spain. It only represents 17 g per day per inhabitant in Portugal and 10.5 g in Japan. Fish may represent an appreciable source of animal protein: 16 g per day per inhabitant in Japan, 13.5 g in Portugal, 12 g in Spain and 7.7 g in Sweden.

As noted above, the rise in the consumption of animal proteins increases the availability of fat from these sources (Tables 2, 4), in particular pork products (pates, meat pies, etc.), milk products, full-cream milk, and 45 or 48% fat cheeses. Accordingly, in countries which have high consumption levels of animal protein, the energy supplied by these animal products is also high (Table 5). When animal protein supplies more than 60% of the total protein, the energy available from the animal products exceeds 1300 Kcal/day, i.e. more than 40% of the total energy available. This is the case in the USA, France, Western Germany, Sweden, Switzerland, etc. It should be noted that butter represents a substantial proportion of this supply, exceeding 200 Kcal/day.

In the Mediterranean countries (Italy, Spain), which traditionally use higher amounts of vegetable oils, the availability of fats of animal origin is only 800 Kcal/day, i.e. 25% of the total available energy.

Trends in the consumption of meat

As shown by the data discussed above, meat represents an essential element of the animal protein supply. The data in Table 6 clearly show the very large differences in meat consumption (expressed in terms of dressed carcass weight) between the various countries and the overall trends over the last 15–20 years. Eastern and Western industrialized countries (about 1.2 billion inhabitants) are consuming high levels of meat (more than 60 Kg per year per inhabitant) and the increase in this consumption has been very marked during the last 20 years (1.5 to 2 Kg increase per year per caput). On the other hand, the poorest countries of Asia (about 1.2 billion inhabitants) only consume 3 Kg of meat per year per inhabitant. Africa (28 million inhabitants) consumes 11 Kg on an average, while the amount of meat available in China (1 billion inhabitants) is about 20 Kg per year per inhabitant (which is due to serious efforts made during the last few years to increase the amount of meat

Table 6. Meat consumption (10^6 metric tons)[a]

Population (10^6)	Country	Years	Buffalo, beef	Mutton, goat	Pork	Poultry	Total	kg per capita
	World							
3290		1961–1965	31.90	6.60	31.0	11.70	83.60	25.40
4043		1976	47.10	7.18	42.0	23.10	121.65	30.00
	Developed countries							
690	Total	1961–1965	16.50	2.33	13.50	7.00	40.50	58.70
762		1976	24.50	2.37	17.90	13.70	59.60	78.20
	USA							
194		1961–1965	8.10		5.40	4.50	18.50	95.30
215		1976	12.20		5.60	7.10	25.40	118.10
	Centrally planned countries							
333	Total	1961–1965	4.75	1.24	7.20	1.20	14.70	43.90
366		1976	8.60	1.20	11.20	2.70	24.10	65.80
	USSR							
230		1961–1965	3.50		3.80	0.75	9.30	40.40
256		1976	6.40		5.70	1.50	15.00	58.60
	Asia							
764		1961–1965	2.03	0.64	8.00	2.00	12.70	16.60
923		1976	2.30	0.75	10.80	3.40	17.10	18.50
1020		1979						21.00
	Developing countries							
1500	Total	1961–1965	8.70	2.37	2.30	1.40	15.70	10.40
		1976	11.90	2.87	3.40	3.40	22.50	11.30
	Africa							
246		1961–1965	1.40	0.63	0.13	0.26	2.80	11.40
328		1976	1.80	0.75	0.21	0.42	3.60	11.00
	Latin America							
247		1961–1965	5.70	0.41	1.43	1.50	8.40	34.00
333		1976	7.97	0.41	2.00	1.60	12.20	36.60
	Asia							
854		1961–1965	0.96	0.51	0.81	0.35	2.70	3.20
1125		1979						3.70

[a]Source: FAO. [2]; dressed carcass values.

available). For half of the world population (26 billion inhabitants), however, the increase in meat consumption has not exceeded 0.5 to 2 Kg per inhabitant during the last 15 years.

A survey of the meat consumption in the industrialized countries since 1955 (Tables 7, 8) shows a steady increase of more than 1.5 Kg per year per inhabitant in almost all of these countries. In some countries, such as the USA, the meat consumption started increasing before World War II; in other countries (France and other EEC countries), increases started in 1950 and in Japan, Portugal, Spain and most East European countries, only in 1960. Some of the richer countries with different food habits have increased their consumption more recently: Switzerland in 1963, Sweden and Denmark in

Table 7. Consumption of dressed carcass (kg/year per capita)[a]

Population (10^6)	Country	Year	Beef	Pork	Horse, mutton, goat	Poultry	Rabbit and other meats	Offal	Total
	OECD								
		1955	21.1	18.0		5.4			54.8
		1959	22.4	20.0		7.4	1.1	3.5	59.0
682.0		1966	26.5	20.7	4.0	11.1	1.3	3.9	67.0
726.0		1972	28.6	24.6	4.0	13.9	1.4	4.1	76.7
751.0		1976	31.8	24.3	3.4	14.9	1.4	4.5	79.7
		1978	30.4	25.9	3.3	16.1	1.4	4.5	81.7
	USA								
		1955	35.0	28.2	2.1	12.5	1.2	5.0	84.0
		1959	39.5	30.7	2.1	15.9	1.2	4.6	94.0
193.0		1966	49.3	26.4	1.8	19.9	1.1	4.7	103.2
206.0		1972	53.6	30.6	1.5	23.6	1.1	4.8	115.2
213.0		1976	60.5	27.0	0.9	23.8	1.1	5.0	118.1
		1978	55.9	28.0	0.7	25.9	1.1	4.7	116.1
	EEC								
		1955	18.0						50.0
		1959	19.9	23.6		4.8	1.0	4.2	59.0
245.0		1966	23.4	26.9	4.2	8.5	1.6	4.9	68.1
255.0		1972	24.4	31.9	4.2	11.4	2.0	5.3	79.3
259.0		1976	25.2	33.0	3.9	12.3	2.2	5.0	82.8
		1978	25.7	35.8	4.0	13.4	2.3	5.2	87.3
		1985	26.8	37.5		14.0			90.5
	France								
		1955	22.3	21.7		7.0	1.8	6.9	71.4
		1959	26.7	26.2		8.1	2.4	7.0	75.1
49.1		1966	29.2	28.7	4.7	11.0	2.4	8.1	86.1
51.7		1972	28.5	32.7	5.0	13.2	6.1	8.8	94.3
53.0		1976	30.8	34.8	5.5	14.7	6.0	9.9	101.7
		1978	32.2	37.2	5.6	15.9	5.7	11.1	107.7
		1985	34.35	41.0	6.0	17 à 18	5.5	10.0	
	Japan								
		1955	1.3	1.5		0.4	1.2	0.5	0.5
		1959	1.6	2.0		0.4	1.6	0.6	6.6
99.0		1966	1.7	5.7	1.3	2.7	2.0	1.2	14.6
107.0		1972	3.5	8.9	1.9	6.3	1.2	1.7	23.5
113.0		1976	3.5	10.7	1.7	7.1	0.7	2.0	26.3
		1978	4.4	12.1	1.8	1.0	0.5	2.3	29.1
	Austrialia and New-Zealand								
		1959							110.0
14.4		1966	40.6	11.3	37.8	6.7		4.9	106.0
16.0		1972	40.5	14.7	35.3	12.1		5.7	111.0
17.0		1976	64.7	11.4	24.8	14.6		5.8	120.0
		1978	60.0	13.0	22.0	17.6		5.7	116.0

[a] Source: OECD. [3, 8]; 1985 Values are projected.

1972 (Table 8). Today, a little more than 1/4 of humanity (1.2 billion inhabitants) is involved in increasing in their consumption of meat.

In the industrialized countries, the average increase in meat consumption exceeds 2.3 million tons per year out of a total increase per year in the world

Table 8. Consumption of dressed meat (kg/year per capita)[a]

Population (10^6)	Country	Year	Beef	Pork	Horse, mutton, goat	Poultry	Rabbit and other meats	Offal	Total
	Denmark								
4.5		1955	21.0	33.0	1.0	2.9	0.7	3.9	62.5
4.6		1963	19.5	31.4	0.4	3.4	0.9	6.0	61.6
4.8		1966	19.0	32.3	0.6	4.0	0.8	6.7	63.4
5.0		1972	19.0	34.1	0.6	5.4	0.6	6.6	63.7
5.07		1976	16.0	39.6	0.6	8.3	0.6	6.5	71.6
		1978	16.8	45.6	1.0	8.4	0.6	7.1	79.5
	Western Germany								
52.7		1955	18.0	33.5	0.8	2.6	0.5	4.0	59.4
56.7		1963	22.4	37.2	0.5	5.0	0.6	4.5	70.2
59.3		1966	22.4	39.8	0.5	6.6	0.7	4.5	74.5
61.6		1972	24.5	49.0	0.3	8.9	0.9	4.5	88.0
61.5		1976	24.0	52.0	0.7	9.0	1.1	5.3	92.0
60.9		1978	24.1	55.7	0.9	9.9	1.1	5.4	97.1
	Great Britain								
51.4		1955	23.0	20.5	10.8	3.9	0.5	3.9	62.6
53.2		1963	26.9	25.7	11.0	6.6	0.5	4.7	75.4
54.8		1966	22.8	26.9	10.9	7.9	0.8	4.8	74.1
55.8		1972	23.0	28.0	9.9	11.5	0.7	4.4	77.5
55.9		1976	24.0	23.5	8.9	11.6	0.1	4.7	72.0
		1978	24.5	25.5	7.2	13.0	0.1	4.8	75.1
	Italy								
48.7		1955	11.1	7.5	1.6	3.0	1.3	1.4	25.8
50.0		1963	16.2	8.5	1.8	6.4	1.2	1.8	35.9
52.0		1966	21.0	10.3	1.7	10.0	1.5	2.1	47.5
53.0		1972	26.0	14.7	2.0	12.0	2.5	3.0	61.2
55.8		1976	23.5	18.8	2.0	16.4	3.0	3.2	67.2
56.4		1978	24.0	21.3	2.4	17.3	3.3	3.3	71.4
	Sweden								
7.3		1955	19.0	24.8	1.2	1.8	1.0	2.8	51.6
7.5		1963	20.0	24.7	1.6	1.8	1.1	3.2	52.4
7.8		1966	19.1	25.2	1.6	2.6	1.0	2.8	52.3
8.1		1972	15.3	28.1	1.2	3.9	1.1	2.5	52.1
8.2		1976	20.3	33.6	1.3	4.8	1.2	2.9	63.9
		1978	18.0	33.9	1.2	5.1	1.6	2.7	62.5
	Switzerland								
5.1		1955	21.2	22.5	1.4	2.5	0.6	3.9	52.1
5.6		1963	24.7	26.4	1.6	5.4	0.7	4.1	62.9
6.2		1966	24.5	28.3	1.5	6.0	1.0	4.3	65.6
6.5		1972	26.3	35.5	1.7	7.0	1.2	4.8	76.5
6.4		1976	25.5	38.7	1.7	7.0	1.2	5.4	79.5
		1978	26.1	41.2	1.8	6.8	1.4	5.6	83.0
	Spain								
29.5		1955	5.3	7.2	3.1	2.7	0.7	1.2	20.2
30.8		1963	8.2	10.0	4.2	6.0	0.5	2.0	30.9
32.3		1966	8.9	12.1	4.5	9.6	0.6	2.2	36.5
33.8		1972	11.3	16.7	4.4	16.4	1.2	2.8	51.6
36.0		1976	13.8	19.6	4.5	19.6	2.8	3.8	64.0
		1978	12.7	22.7	4.0	20.7	3.3	3.7	67.2

[a]Source: OECD. [3, 8].

Table 9. Consumption trends (per caput) of various animal products[a]

Country	Year	Fresh milk and other fresh diary products (kg/year)	Cheese (kg/year)	Eggs (kg/year)	Fish and shell fish (kg/year)
OECD					
	1955				
	1963	113.1	5.5	13.3	
	1966	112.9	6.0	13.6	
	1972	108.1	7.5	14.8	
	1976	105.4	8.4	14.1	
USA					
	1955	175.0	5.5	20.8	5.8
	1963	163.6	6.2	18.9	6.1
	1966	159.4	6.5	18.5	6.3
	1972	143.3	8.4	18.2	6.5
	1976	135.5	9.3	16.4	6.9
EEC					
	1955				
	1963	107.2	7.7	12.3	
	1966	107.8	8.4	12.9	
	1972	104.9	10.2	13.9	
	1976	103.0	11.4	13.6	
France					
	1955	90.0	7.8	10.7	10.4
	1963	85.0	10.4	10.7	16.1
	1966	83.3	12.0	11.2	17.0
	1972	74.5	14.5	12.5	17.0
	1976	76.2	16.2	13.0	18.7
Japan					
	1955	8.0	0.0	4.3	45.2
	1963	15.8	0.2	10.2	36.5
	1966	21.2	0.5	11.8	30.0
	1972	27.0	0.7	16.2	34.0
	1976	30.1	1.0	16.0	37.7
Australia and New-Zealand					
	1955	140.0	3.0	12.0	
	1963	146.0	3.4	12.8	
	1966	144.0	3.4	13.0	6.5
	1972	140.0	4.6	13.0	5.6
	1976	130.0	5.3	13.0	7.2

[a]Source: FAO. [2]; OECD. [3, 4].

of 3.5 million tons. One fourth of humanity profits from two thirds of the annual increase in the availability of meat: the industrialized countries consume 90 million tons of meat out of a total of 138 million tons, i.e. 65% of the overall world consumption (Table 11).

Maximum meat consumption seems to range around values equivalent to 115 to 120 Kg of carcass meat per year which corresponds to the consumption present in the USA since 1972 and in Australia and New Zealand since 1975. The richest countries (the USA, New Zealand and Australia) prefer red meat (beef and/or mutton). However, substitutions may occur when total consumption reaches 70 Kg per year per inhabitant. Thus, Great Britain, whose

Table 10. Consumption trends (per caput) of various animal products[a]

Country	Year	Fresh milk and other fresh dairy products (kg/year)	Cheese (kg/year)	Eggs (kg/year)	Fish and shell fish (kg/year)
Denmark					
	1955	175.0	7.1	10.0	13.6
	1963	177.0	9.0	12.5	15.4
	1966	173.9	9.0	12.3	24.9
	1972	152.8	10.8	10.7	35.0
	1976	144.0	9.1	11.6	35.0
Western Germany					
	1955	119.0	6.5	11.4	11.0
	1963	100.9	7.7	13.0	8.6
	1966	97.8	8.9	14.1	7.4
	1972	98.2	11.0	17.3	7.3
	1976	8.9	12.3	16.8	9.0
Great Britain					
	1955	163.0	4.3	13.4	10.0
	1963	157.0	4.7	15.6	9.6
	1966	159.3	4.8	16.1	9.3
	1972	157.3	5.4	16.0	8.3
	1976	164.0	6.1	14.5	7.9
Spain					
	1955	67.0	1.4	5.2	22.6
	1963	70.2	1.6	10.0	28.6
	1966	70.0	2.0	11.8	28.6
	1972	80.4	2.9	16.4	28.0
	1976	105.2	3.6	17.0	32.0
Italy					
	1955	57.0	8.2	8.1	6.8
	1963	61.1	8.6	9.3	7.9
	1966	68.1	9.3	9.6	10.2
	1972	72.6	10.9	10.7	10.0
	1976	72.6	12.4	11.1	10.8
Sweden					
	1955	193.0	7.5	11.7	18.2
	1963	178.2	7.7	11.8	19.9
	1966	173.3	8.1	11.9	19.8
	1972	154.3	9.0	12.6	20.6
	1976	160.2	11.9	12.3	23.5
Switzerland					
	1955	210.0	8.1	9.7	3.0
	1963	190.5	8.3	9.7	3.9
	1966	185.1	8.8	10.2	4.3
	1972	172.2	10.7	11.3	5.0
	1976	170.2	12.0	11.0	4.6

[a] Source: FAO. [2]; OECD. [3, 4].

standard of living has increased little since 1965, replaces beef and mutton (more expensive production) by pork and poultry.

In the coming ten years, an increase in meat consumption of more than 2 million tons per year can be expected in the industrialized countries (1.2 billion inhabitants). In China it will probably be around 1 million tons, and

Table 11. Production of protein-rich products (10^6 tons/year)[a]

Products	Years	World (4.335)[b]	Western Europe (370)	Eastern Europe (375)	North America (244)	Developing Asia (1.215)	Centrally planned Asia (1.024)
Meat	1961–65	84.25	16.52	14.60	20.10	2.87	12.82
	1969	102.24	20.15	18.25	23.80	3.47	14.80
	1972	113.44	22.18	21.20	25.60	3.77	18.10
	1976	126.51	25.08	22.30	27.60	4.18	20.60
	1978	134.67	26.60	25.10	27.80	4.56	23.60
Milk	1961–65	358.00	111.42	93.50	65.30	28.90	4.70
	1969	397.70	119.44	115.50	61.10	31.90	5.10
	1972	411.60	122.60	119.20	62.50	34.00	5.60
	1976	437.80	129.20	127.70	62.20	36.90	6.70
	1978	457.30	135.60	135.20	62.90	38.20	7.30
Eggs	1961–65	16.72	3.74	2.63	4.10	0.53	2.80
	1969	20.46	4.48	3.33	4.30	0.76	3.50
	1972	22.56	4.92	4.10	4.40	0.91	3.60
	1976	24.41	5.05	4.76	4.10	1.10	4.00
	1978	26.22	5.20	5.38	4.30	1.24	4.40
Fresh water fish	1961–65	5.61	0.15	0.79	0.37	1.87	1.17
	1969	6.78	0.16	1.04	0.32	2.40	1.19
	1972	7.09	0.16	1.17	0.32	2.38	1.16
	1976	7.51	0.18	1.34	0.33	2.50	1.34
	1978	7.74	0.18	1.04	0.39	2.58	1.34
Salt water fish	1961–65	35.84	7.95	3.67	2.60	2.90	3.01
	1969	48.28	9.38	6.28	2.54	4.74	3.73
	1972	49.30	10.01	7.60	2.49	5.64	4.76
	1976	55.71	10.89	10.33	2.60	7.02	5.49
	1978	55.22	10.22	8.77	2.97	7.97	5.52
Shellfish	1961–65	4.40	0.63	0.11	0.98	0.51	0.51
	1969	5.31	0.80	0.12	0.98	0.84	0.59
	1972	5.97	0.96	0.20	1.02	1.13	0.71
	1976	7.01	0.96	0.11	1.16	1.68	0.99
	1978	7.75	0.96	0.22	1.34	1.89	1.11
Fish and shellfish	1974	69.50	1.40	10.60	4.00	10.80	8.20
	1976	70.23	12.10	11.80	4.10	11.20	7.80
	1978	70.70	11.40	10.00	4.70	12.40	8.00
			Developed countries			Developing countries	
Edible Fish	1974	48.90	25.10			23.80	
	1976	50.70	25.60			25.10	
	1978	52.70	24.90			27.80	
Non-edible fish	1974	20.60	13.40			7.20	
	1976	22.10	14.40			7.70	
	1978	21.00	13.60			7.40	

[a]Source: FAO [2].
[b]1979 Population in millions.

for the rest of the world about 2 million tons, taking into account the very large increase in the population density of these countries and the proportion of these countries becoming more industriablized. The total increase, in the coming 10 years, will be about 5 million tons.

Trends in the consumption of other animal protein sources

Eggs

In most industrialized countries, the per caput consumption of eggs has progressively increased and is now about 14.5 Kg per year (Tables 9, 10). In the Eastern and Western industrialized countries, about 250 eggs per year are consumed per inhabitant. One fourth of humanity, i.e. the industrialized countries, produces and consumes two thirds of the eggs, 17.5 out of 26.5 million tons.

Milk

In industrialized countries, milk consumption strongly depends on food habits (Tables 9, 10). Some countries are traditionally consumers of pasteurized milk (160–180 Kg): Sweden, Switzerland, Great Britain, Denmark and the USA. In most of these countries, milk consumption is decreasing. Within the EEC, annual per capita milk consumption is almost constant (107–103 Kg), but it has slightly decreased within the OECD (from 113 to 105 Kg in 15 years). In the industrialized countries of Eastern Europe, milk consumption remains high: 120–160 Kg per year per inhabitant. According to the data for the last 15 years, the industrialized countries produce and consume 85% of the world's milk.

Milk production in the industrialized countries of Eastern and Western Europe is increasing owing to the higher productivity of the dairy cows. In France, 24 million tons of milk are collected and, within the EEC, 100 million tons. These amounts will continue to increase if no efficient means of control are used.

In tropical regions, the development of milk production seems to be difficult because the climate is not well fitted for high yielding dairy cows. However, buffalo cows, producing 1000 Kg milk per lactation (7% fat), as well as goats, may be a solution for obtaining a supply of milk for children in these developing countries.

Cheese

In most industrialized countries, cheese consumption regularly increases by 200 to 400 g per year per inhabitant. Cheese consumption is steadily increasing in Eastern Europe, from 5 Kg per year per inhabitant to 7 Kg (more than 200 g per year per inhabitant), during the period 1970–1979. The industrialized countries (1.2 billion inhabitants) produce and consume more and more cheese (83% of the world production in 1970, i.e. 6.4 million tons out of 7.7; 86% in 1979, i.e. 9.5 out of 11 million tons). Some South American countries have started cheese maufacturing and consume about 1–2 Kg per year per caput.

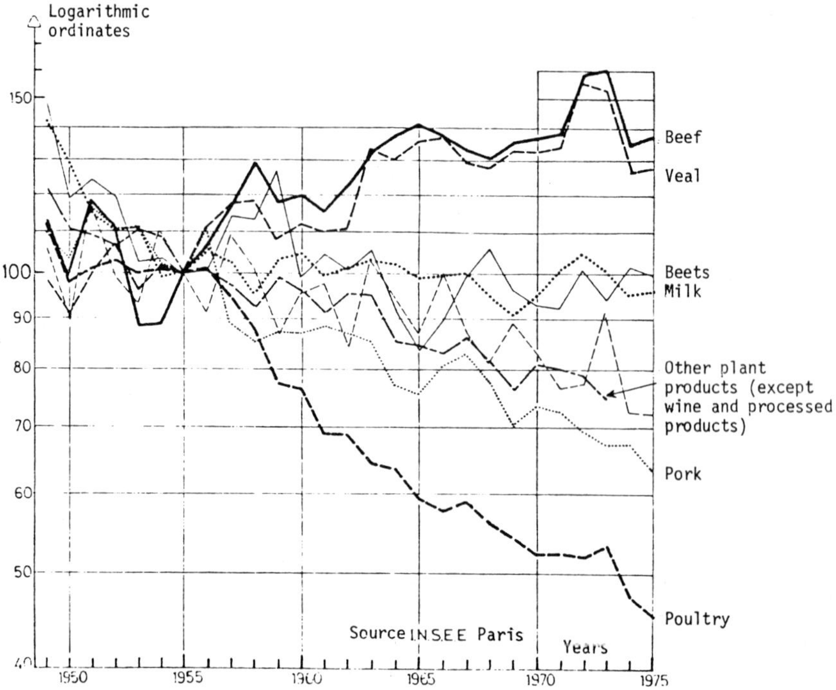

Figure 1. Evolution of agricultural production prices (in constant French Francs; 1955 = 100).

Future animal production trends

Animal production is increasing steadily in all of the industrialized countries which possess an efficient agriculture. The production costs (Figure 1) indicate the situation in France during the last 20 years. The evolution of prices indicate that the productivity of poultry has increased most (in 1955, chicken and beef prices were similar), that of pigs less and that of beef cattle even less. Other factors such as organization of markets and economic and political impact of professional associations have reduced the lowering of prices of some products (guaranteed prices), both in France and within the EEC, such as cereals, sugar, etc.

A better control of aquaculture and aviculture (increasing by 5% per year in the world) in the industrialized countries might lead to a greater increase in production. Aquaculture and aviculture are economic models that the industrialized countries may export to the developing countries, which rightfully desire to increase their consumption of animal protein.

World protein balance

The world protein balance within the overall agriculture and food sector is as follows: 180 million tons of proteins are produced in the form of plant

Table 12. Japanese consumption of industrially processed plant protein (tons/year)[a]

	Products	Defatted soybean mean equivalent
Unfermented	Tofu	84,200
	Kori-tofu	22,400
	Other products (texturized protein included)	87,800
Fermented	Shoyi (souce)	173,000
	Niso (soyabean paste)	157,400
	Natto (soyabean seeds)	55,200
	Total	580,000

[a]Source: Dronne [1].

products: cereals, legume seeds, oil-seeds, etc. Of this protein, 80% can be used directly by man so that if it were equitably distributed, it would supply 90 g of protein to each inhabitant in the world. But, at the present time, more than half (100 million tons) of this protein is used for animal feeding in the developed countries. These 100 million tons are composed of 65 million tons of cereals and 35 million tons of other protein-rich products. All of these products are transformed through animal feeding into 35 million tons of animal protein (plus 5 million tons of fish protein); this represents a yield of 35–40% relative to the concentrates. If we add the forages and forage derivatives consumed (estimated to be about 220 million tons of protein), the yield is only about 12%. The 35 million tons of animal proteins (non-fish) are not equally distributed, since the developed countries consume about 21 million tons (for 1.2 billion inhabitants), China about 4 million tons (1 billion inhabitants) and the rest of the world (2 billion inhabitants) about 9 million tons (Table 11). With respect to the 5 million tons of fish protein, 2.5 million tons are consumed in the developed countries and 2.5 million tons in the rest of the world (3.2 billion inhabitants).

The 80 million tons of plant proteins not consumed by animals do not represent a sufficient quantity to satisfy the human requirements; only some 88% (70 million tons) can be used directly by man. This is only 45 g per day per inhabitant. However, the distribution of plant proteins is more equal than that of animal proteins.

'By-passing' the animal consumption of purified plant proteins

Plant proteins are not greatly appreciated by the population of the industrialized countries because they are diluted, more or less, by large amounts of various carbohydrates. In fact, protein only constitutes about 8% of cereal meals and 20% of beans or peas. Sedentary people have low energy requirements (2400 Kcal per day for men and 2000 Kal per day for adult women) and desire to feel comfortable during the post-prandial period. Further, some of the carbohydrates present may cause flatulence.

Table 13. Japanese consumption of new 'purified' protein (tons/year)[a]

Years	Total	Soyabean	Gluten	Powder	Texturized
1971	27,300	15,900	11,400	17,400	9,600
1973	35,600	12,800	22,800	17,900	17,600
1975	37,000	13,300	23,700	16,300	20,000
1976	41,100	15,800	25,300	18,500	22,600
1977	44,000	17,000	27,000	19,600	24,400

1977:	Soyabean	Gluten
Powder products (total)	9,500	10,100
– Isolates	6,000	
– Concentrates	3,500	
Texturized products (total)	7,500	16,900
– Dry extruded	6,000	200
– Moist	1,500	12,000
– Pasty		4,700

[a]Source: Dronne [1].

Even populations in countries in which legume proteins (home-prepared) are traditionally consumed have difficulty in maintaining their consumption levels despite an increasing industrialization. Japan still consumes 800 000 tons of soybean meal of which 580 000 tons are industrially processed into traditional products (Table 12). Soybeans provide each Japanese with 3.7 Kg protein per year (i.e. 10g per day). Consumption, however, is tending to decrease and the development of new protein sources is slow: texturized gluten proteins seems to be the most promising product (Table 13).

In Europe, the manufacturing and marketing of third generation products (improved and texturized concentrates and isolates) are increasing, especially in Great Britain. In the USA, half of the concentrated protein is consumed by domestic animals; the fraction consumed by man constitutes an amount which is less than 1% of the consumption level of the meat proteins.

Conclusions

High meat consumption in man depends on various factors: socio-cultural (standard of living), emotional (red meat), physiological (comfort, dietetics, health), and organoleptic. Meat is appreciated because of its juiciness, succulence, original texture and flavor. The physiological factors involved in meat consumption should be better analysed so as to develop sufficiently attractive meat substitutes (purified and texturized plant proteins). These products should be adapted to the various sociological habits of different populations.

Acknowledgment. Grateful acknowledgment is made to Mrs. Rerat for her translation of this paper.

References

1. Dronne Y Le probleme mondial du soja, Institute National de la Recherche Agronomique, Paris. 1981.
2. Fao (Food and Agriculture Organization), Production Year Book 33, (1979) FAO, Rome.
3. OECD (Organization For Economic Cooperation and Development). Food Consumption Statistics, 1955–1971, OECD, Paris, 1973.
4. OECD (Organization For Economic Cooperation and Development). Food Consumption Statistics, 1969–1976, OECD, Paris, 1978.
5. OECD (Organization for Economic Cooperation and Development). Milk, Milk Products and Egg balances in OECD, 1964–1977, OECD, Paris, May, 1979.
6. OECD (Organization For Economic Cooperation and Development). Milk, Milk Products and Egg Balances in OECD, 1973--1978, OECD, Paris, May, 1980.
7. OECD (Organization For Economic Cooperation and Development). Meat Balances in OECD Member Countries, 1963–1976, OECD, Paris, May, 1978.
8. OECD (Organization For Economic Cooperation and Development). Meat Balances in OECD Member Countries, 1971–1973 OECD, Paris, April, 1980.
9. USDA (United States Department of Agriculture). Foreign Agriculture Circular. Grains, Oil-seeds and Products, 1980–1981. U.S. Government Printing Office, Washington, D.C. 1982.
10. SCESS: Service Central des Enquetes et Etudes Statistiques (Ministère de l'agriculture), Head Division of Statistical Surveys and Studies (Ministry of Agriculture).

Crude protein content and aminoacid composition of seeds: variability and correlations

J. MOSSÉ and J. BAUDET

Laboratoire d'Etude des Protéines, Département de Physiologie et Biochimie Végétales, I.N.R.A. 78000 Versailles, France

Key words: cereals, legumes, oilseeds, seed proteins, amino acid composition

Abstract. The main crop species (cereals, legume seeds and oilseeds) are reviewed with respect to their seed proteins. In the first section, a few points dealing with available analytical methods and nitrogen conversion factors are briefly discussed. The average crude protein contents of different cultivated species of grain are compared. The variability of seed protein content within a same species (phenotypic variations and biochemical mutations) is also discussed. The same kind of survey is then devoted to seed aminoacid composition in a second section. The analytical problems which are involved are summarized. The average aminoacid compositions of the main species are compared and the variations occurring within the same species are described, for the phenotypic ones as well as for some mutants. A last section concerns the relationships between aminoacid composition and protein content. The significance of these relations and their practical consequences are indicated.

Introduction

Seed protein content and aminoacid composition are known to be different from one genus to the other, from one species to the other and even from one cultivar to the other. But they can also differ between plants of the same variety according to growth and development circumstances, or even between individual seeds according to their positions along the stem or twigs of a single plant. Such differences arise from two main groups of factors: genetic and varietal factors on the one hand; environmental influences (including agricultural practices and freedom from infestation) and physiological conditions, on the other hand. Both kinds of factors are almost always involved together, they almost always interact one on the other, and there is a high degree of overlapping which prevents a clear distinction among their respective influences.

Moreover, analytical determinations of protein content and aminoacid composition are generally made with samples resulting from milling of several hundred or thousand seeds. The requirement of a great number of seeds can be firstly explained by the small size of most seeds, the weight of each seed may only be ten to hundred mg (in the past, analytical techniques have not been adapted for analyses of very small quantities of flour). It can also be explained by the fact that, from a practical point of view, in order to be representative, samples must correspond to a representative population of numerous seeds. The consequence is that analytical data are made of averages between which possible differences are already buffered.

Nevertheless, if a clear-cut separation between the two kinds of factors remains difficult, it is obvious that the influence of genetic factors is generally even more difficult to be accurately determined. It is possible to investigate the influence of physiological conditions by comparing seeds arising from the same line according to environmental factors. In other words, it is possible to preclude, or at least to lower tremendously, any genetic influence by working with seed samples having the same genetic background. But the reverse appears as much more difficult: when seed samples from distinct lines or cultivars are compared, even if environmental conditions are identical, the latter ones may be unequally propitious to an optimal expression of the genetic background according to varieties.

This discussion will be concerned with the main crop species, with a particular consideration of legume seeds (already reviewed elsewhere, see Mossé and Pernollet [41]) because of their diversity which is itself related to the exceptional magnitude of the botanic family of leguminosae. The first two Sections deal respectively with seed protein content and aminoacid composition. After a few analytical considerations, the most frequent values found among the main crop species are surveyed with special attention to variability of any origin. In the third Section, the relationships between seed aminoacid composition and protein content are considered.

Seed protein content

Analytical determination

The diversity of proteins is so tremendous that they are among the substances of which any accurate estimation is the most difficult. Each protein species has its own aminoacid composition and its own nitrogen content. In fact, there is no factor of any kind really able to provide any accurate comparison of the amounts of two different single proteins or of two different protein mixtures. That approximations are sometimes underestimated is therefore inevitable.

Analytical methods which are used in order to quantitate the amount of protein can be classified according to the factor or property upon which the determination is based. At least four kinds of methods can be distinguished according to whether the analytical principle is based on a rather specific element (like nitrogen with Kjeldahl, Dumas or proton activation), on a specific kind of chemical bond (like peptide bond with biuret reaction), on a particular property of one or several aminoacid residues (with spectrophotometric or colorimetric methods), or on a combination of the latter (with infrared reflectometry). Several of these methods have been compared for the analyses of cereals [27] and legume seeds [53]. Despite its limitations, nitrogen determination remains the only reference technique: nitrogen content (d.b.; on a dry matter basis) corresponds to a well-defined parameter with both quantitative and additive features. Therefore the following discussion will be almost exclusively restricted to studies involving nitrogen determinations and protein contents will be always calculated by multiplying the nitrogen percentage (d.b.) by the coefficient 6.25. In spite of its conventional feature,[1] the choice of this value results from the fact that many proteins – and particularly protein mixtures – of various origins show a nitrogen content of about 16%, the reverse of which is 6.25. This conversion factor is also the only one used in the field of nutrition science. With seed proteins, it gives in fact their protein content in terms of 'crude protein.' By definition crude protein includes many different nitrogen compounds: free aminoacids either proteic or uncommon, nucleic acids, osamines, complex lipids, alkaloids, puric and pyrimidic bases, nucleotides, and so on. All of these compounds are included in 'non protein nitrogen' (NPN). The literature is unfortunately very poor in quantitative data about the different NPN components. But systematic analyses of NPN content of many seeds [21] show that the ratio of NPN to total seed nitrogen is practically always in the region of 5 to 10% in cereals and of 10 to 15% in legumes. Proteins are therefore highly predominant, making up 85 to 95% of the total seed nitrogen.

As for the seed nitrogen to protein conversion factor, several authors [56, 55, 15, 28, 54, 60] have determined more realistic values than the classical 6.25 factor, by calculating various factors from aminoacid compositions. These values range from 5.3 to about 6 (5.3 for sunflower, 5.46 for peanut, 5.52 for pea, 5.71 for soybean, 5.83 for wheat, barley and rye and 5.95 for rice). It will be seen further that variations in aminoacid composition which result from changes in protein content, make it useless to require a very high accuracy in the nitrogen to protein conversion factor. The 6.25 factor is useful provided that we are always aware of the fact that the utilization of 6.25 as a factor always gives an overestimate, ranging from 5 to 15%, for the total protein content.

Intergeneric or interspecific variations of seed protein content

In Europe, as all around the world, cereal grains are the most important source of proteins although they are generally rich in starch and rather poor in protein. Their protein content ranges from 7 to 18% and, in very extreme conditions from 5 to 23%, as shown for example by the systematic

[1] Theoretically, the nitrogen content of proteins can vary between two extreme values: 8.6% for polytyrosine and 35.9% for polyarginine. Moreover, conjugated proteins can have a low nitrogen content when their prosthetic group is devoid of nitrogen but may be proportionately significant (e.g. some glycoproteins).

screening of the world wheat collection [31]. The averages are a little lower for rice and maize (about 10%) than for other species (12–14%).

The legumes correspond to a large botanical family in which the seeds show a very large variation in protein content: 12 to 55% among 320 different species belonging to 86 genera analysed by Earle and Jones [21]. Typical protein contents found among seeds of the most highly cultivated legumes are indicated on Figure 1. The right side of Figure 1 shows the range of the

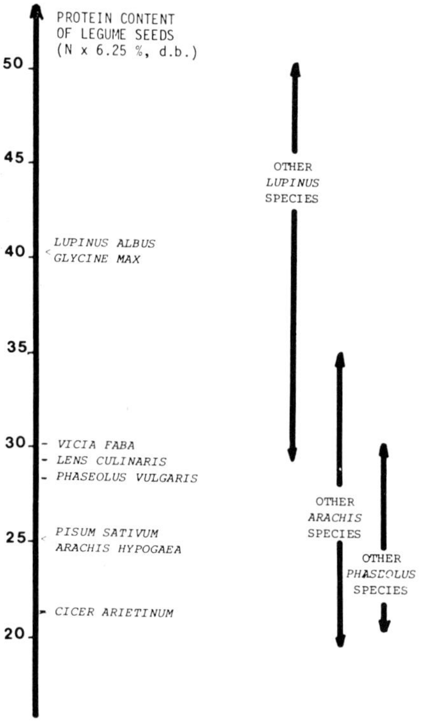

Figure 1. Average protein content (d.b.) of main consumable legume seeds (after Mossé and Pernollet [41]).

most commonly observed protein contents of different species of three genera including *Arachis,* of which 14 different wild species have been studied by Cherry [15] and *Lupinus* which exhibit the highest seed protein content among all cultivated legumes, with protein levels ranging from 27 to 50%. Only very few legume seeds have significant oil contents: about 45% in *Arachis hypogaea,* 20% in *Glycine max* and 5 to 15% in many *Lupinus* species.

Several important oilseeds arise from other very different botanical families: *Brassica napus* and *Helianthus annuus* both show seed protein contents varying from about 15 to 30% with an average of about 22%. But for oilseeds, data in the literature are generally for oilseed cake, the protein content of which depends also upon the seed oil content.

Intraspecific variations of seed protein content

By growing cereals of the same variety in pots with very different fertilizing conditions, Eppendorfer [22, 23, 24] performed an intervarietal study. He obtained protein contents which varied from 6 to 23.8% for oats (var. Selma) and 9.75 to 25.6% for barley (var. Lofa). Practically the same ranges were observed for one wheat and one rye variety.

Different intraspecific studies have dealt with legume seeds. For instance, the protein content of 150 lines of *Cicer arietinum* varied from 15 to 29.6% with an average of 22.2% [53]. Among 87 different samples of *Phaseolus vulgaris* seeds, protein contents varied from 22 to 39% with an average of 28% [21]. A systematic analysis of 50 single seeds randomly picked from a seed sample of the same *Pisum sativum* variety ('Minarette') harvested the same year at the same place showed protein contents ranging from 23 to 40.5% [3]; even when the seeds were obtained from only two single plants, the protein content of single seeds ranged from 21.6 to 32% for one pea plant and from 23 to 34% for the other. The individual seed weight was also taken into account with *Pisum sativum* [58, 59, 3]. The results showed that the seeds were very close for their individual weight and protein content when they were located inside the same pod or in pods bound to the same nod. But both seed weight and protein weight per seed decreased when the distance from the roots increased. Baudet et al. [3] have shown the existence of a linear relationship between seed weight and protein weight per seed (Figure 2). This latter relationship is independent of the seed localization in the plant but may change with the genetic characteristics of the varieties studied.

From the preceding discussion it appears obvious that the numerous possible variations of seed protein content do not decrease very much when restricting the samples from a genus to a species, from a species to a variety and even more from a variety to a single plant.

Seed aminoacid composition

Analytical determinations

There are plenty of data on seed aminoacid composition in the literature, some of which have been pooled by F.A.O. [25]. But even when the same species have been analysed by different laboratories, results are not always in good agreement. A comparison of the composition of the same sample analysed by different laboratories has revealed tremendous discrepancies [1]. In addition to the different causes of variation which have been discussed already about seed protein content, discrepancies can also be partially explained by the complexity and diversity of the problems posed to the chemist.

For the most accurate analyses the following guidelines should be obeyed,

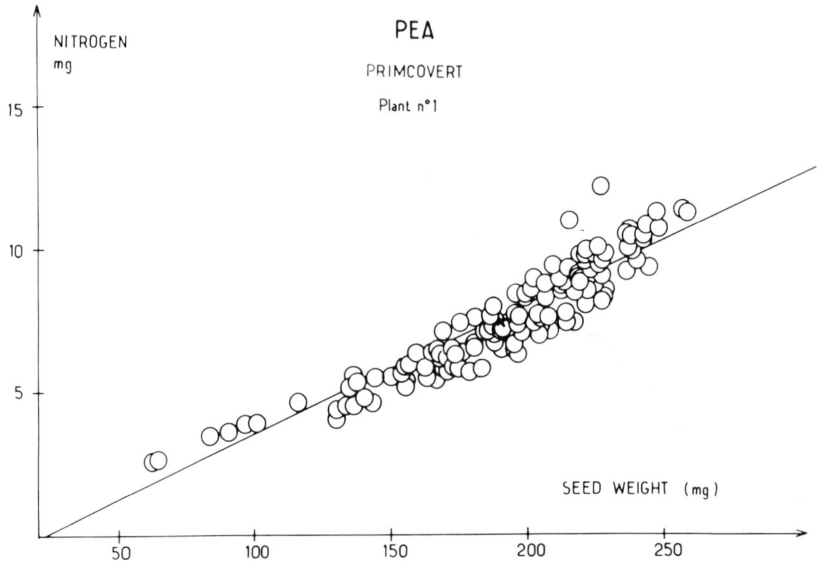

Figure 2. Amount of nitrogen per seed as a function of seed weight for 165 seeds of a pea plant (Variety Primecovert; after Baudet et al. [3]).

some of which have already been mentioned by F.A.O. [25] : (1) The origin and nature of the sample (name of the cultivar) must be specified, as well as its total weight and the average weight of the seeds involved. (2) The protein content of the seed sample must be preliminarily determined. (3) Aminoacids should be analysed after 24 and 48 hour hydrolyses in 6 N HCl at 115°C and subsequent chromatography on ion exchange columns performed with automatic analyzers. Separate analyses are carried out for methionine and cysteine (hydrolyses after performic acid oxidation), and also for tryptophane (alkaline hydrolyses). (4) The amounts of 18 (or 17 if tryptophane is not analysed) of the 20 aminoacids of the code of Nirenberg are tabulated, because of the impossibility to distinguish glutamine from glutamic acid (Gln + Glu = Glx) and asparagine from aspartic acid (Asn + Asp = Asx). (5) The uncommon aminoacids which might occur must be taken into account, whatever they are: the proteic ones resulting from postranslational modifications (like hydroxyproline) as well as the free ones (like canavanine). (6) The results give (or enable calculating) the total weight of aminoacid recovered; or better, they give the percentage of recovered nitrogen, i.e. the yield of the analysis.

Intergeneric and interspecific variations of seed aminoacid composition

By taking into account the preceding analytical considerations, data available in the literature have been selected from Pion and Fauconneau [48] for cereals (Table 1) and for oilseeds (Table 2) and from Mossé and Pernollet [41] for legume seeds (Table 3). Even though cereals consist of a rather homogenous botanical group, legumes comprise a botanical family and oilseed species are of great botanical diversity, comparisons of seed aminoacid content have been made inside each of these three major groups. In each group and for each aminoacid, the ratio max/min, i.e. highest content/lowest content, has been calculated. This allows for roughly classifying aminoacids according to the variability of their content.

Among cereal grains, Table 1 shows that this ratio is very low for Ile, Thr, Ser and Phe (1.22 to 1.26) and less than 1.5 for Val, His and Glx. By way of compensation, Ala, Leu and Arg exhibit the highest ratios (2.13 to 2.4) while all the other amino acids correspond to intermediary ratios (1.5 to 2). Among the latter, it must be noticed that Lys (1.84), Met (1.89), Cys (1.94) and Pro (2.08) content can almost double when passing from the poorest species involved to the richest one. Moreover, for essential aminoacids like lysine and the two sulfur aminoacids, the poorest species is always the same: sorghum whereas the richest species are either oats or rice or both.

Table 2 gives the composition of four oilseeds. Among the 17 aminoacids involved, 13 show ratios (max/min) lower than 1.5. For all aminoacids except one (Lys), the ratio is less than the corresponding one related to cereal species. Though the four oilseed species belong to four very different botanical families (Compositae, Crucifereae, Linaceae and Pedaliaceae), their seeds are curiously less different from one another in their aminoacid compositions than are the seeds of the grains of cereals which belong to the same family. Also, three aminoacids show the highest ratios (Lys: 2.12; Arg: 2.04; Pro: 1.69) just as they do among the cereals.

Table 3 shows that legume seeds can differ much more than cereals or oilseeds in their aminoacid compositions. For the relevant species, max/min ratios are very low for Leu, Ser, His and Asx and less than 1.5 for Ala, Val and Glx. Among these 7 aminoacids, 5 are common to the group of the 7 aminoacids which also show a ratio of less than 1.5 in cereal grains. On the contrary, Met (2.29) and Lys (2.12) exhibit the highest ratios, while the 8 other amino acids have ratios ranging from 1.55 to 1.91.

If the different species of the three groups (cereals, oilseeds species and legumes) are now compared together, it is interesting to note that 6 of the 17 aminoacids involved show high ratios practically in every group, which means that they exhibit significant interspecific variations in their content: they are Lys, Met, Cys, Tyr, Pro and Arg. This could be explained by the differences in composition of the main storage proteins of the seeds. But if taking into account the ratio max/min allows rough comparisons, this explanation is of course not sufficient. Ranges between the highest and the lowest content of one aminoacid are sometimes quite different. For example, if lysine exhibits the same ratio (2.12) among the four species of oilseeds of Table 2, its content varies from 2.5 g per 16 g of nitrogen in sesame to 5.3

Table 1. Amino acid composition (g/16 g N) of cereal grains[a]

Species:	Triticum aestiv.	Hordeum vulg.	Secale cer.	Avena sat.	Oryza sat.	Zea mays	Sorghum vulg.	Pennisetum typh.	ratio (max/min)
Gly	4.3	4.3	4.3	5.1	4.8	4	3.3	3.9	1.55
Ala	4	4.3	4.4	4.9	6.05	7.7	9.6	7.7	2.4
Val	5.1	5.3	5.15	5.8	6.9	5.4	5.65	5.7	1.35
Leu	6.9	6.9	6.3	7.35	8.85	12.3	13.4	9.4	2.13
Ile	3.8	3.9	3.8	4.2	4.65	3.95	4.45	4.2	1.22
Ser	4.7	4.2	4.5	4.8	5.3	5	4.65	4.7	1.26
Thr	3.2	3.5	3.4	3.5	3.7	3.8	3.4	4	1.25
Tyr	3.2	3.3	2.8	4.1	5.1	4.25	4.5	3.4	1.82
Phe	4.6	5.1	4.5	5.1	5.7	5.1	5.6	4.8	1.24
Pro	9.1	10.1	8.65	5.3	4.85	8.85	8.5	6.1	2.08
Met	1.45	1.55	1.65	1.6	2.65	2.2	1.4	2.5	1.89
Cys	2.6	2.6	2.5	3.5	2.6	2.5	1.8	2.5	1.94
Lys	3.35	4	4.1	4.25	4.25	3.25	2.3	3.7	1.85
His	2.5	2.15	2.25	2.2	2.5	2.85	2.3	2.6	1.33
Arg	5.3	5.2	5.5	6.8	8.8	4.8	3.9	6	2.26
Asx	5.3	5.9	7.4	7.8	9.4	6.65	6.8	8.2	1.77
Glx	26.4	21.5	21.7	19.8	18.45	18.55	20.9	18.1	1.46
Total	95.8	93.8	92.9	96.1	104.55	101.15	102.45	97.5	
Protein (%)	9.3	9.4	11.4	9.7	8	11	10.2	9.4	

[a]After Pion and Fauconneau [48]; for each amino acid the maximum and minimum contents are underlined with solid and dashed lines, respectively; protein = N × 6.25.

Table 2. Amino acid composition (g/16 g N) of several oilseeds[a]

Species	Helianthus annuus	Brassica napus	Linum usitatissimum	Sesamum indicum	ratio (max/min)
Gly	5.7	5.05	6.15	4.8	1.28
Ala	4.3	4.35	4.75	4.6	1.1
Val	5.7	5.5	5.5	5.1	1.12
Leu	6.35	6.8	6.1	6.6	1.11
Ile	4.65	4.2	4.6	4	1.16
Ser	4.35	4.35	5.1	4.65	1.17
Thr	3.65	4.4	3.9	3.5	1.26
Tyr	2.8	3.3	2.65	3.75	1.42
Phe	4.8	4.1	5.05	4.6	1.23
Pro	4.2	6.1	3.85	3.6	1.69
Met	2.2	2	1.9	2.9	1.53
Cys	1.95	2.7	2.1	2.3	1.38
Lys	3.6	5.3	3.9	2.5	2.12
His	2.5	2.7	2.1	2.4	1.29
Arg	8.5	6	9.6	12.25	2.04
Asx	8.9	6.8	9.9	8.15	1.46
Glx	21.05	17.35	20.9	19.6	1.21
Total	95.2	91	98.05	95.3	

[a]After Pion and Fauconneau, [48]; for each amino acid the maximum and minimum contents are underlined with solid and dashed lines, respectively.

in rapeseed whereas among all the legume seeds of Table 3, lysine content changes from 3.45 per 16 g of nitrogen to 7.3 in Lentil, and reaches almost the same high level (7.2) in chickpea and in garden pea. Such differences are well known.

Intraspecific variations of seed aminoacid composition

The question to be discussed now is that of intraspecific variations of seed aminoacid composition. At first sight, the only cases where significant changes have been pointed out are those of biochemical mutants [43, 45]. As far as we know, until now they have been observed only among cereals.

The first mutants were detected in maize by Mertz et al. [35] and Nelson et al. [46]. The biochemical changes which are involved result from the presence of recessive single genes like *opaque-2* or *floury-2*. They have been shown to increase considerably the lysine content of the proteins of maize endosperm (by up to two or three times the content of normal endosperm). As for the whole grain, the lysine content of its grain protein is significantly increased and sometimes doubled. With maize hybrid 'INRA 260', in which *opaque-2* or *floury-2* genes had been introduced, Mossé et al. [40] have found 3.7 g of lysine (per 16 g of nitrogen) in *opaque-2* and 3.0 g of lysine *floury-2* instead of the 2.5 observed in normal grains which were practically isogenic compared with the two mutant samples except for the mutant locus. Several other maize mutants of the same kind have been found subsequently (see [44] and [51] for reviews).

Table 3. Amino acid composition (g/16 g N) of legume seeds

Species:	Arachis hypo.	Cicer ariet.	Glycine max	Lens culin.	Lupinus alb.	Phaseolus vulg.	Pisum sativ.	Vicia faba	ratio (max/min)
Reference:	(A)	(B)	(A)	(C)	(D)	(D)	(D)	(E)	
Gly	5.9	4	4.4	3.9	3.85	3.8	4.4	4.3	1.55
Ala	4.05	4.1	4.3	4	3.1	4	4.35	4.15	1.40
Val	4.4	4.6	5.15	4.2	3.9	4.85	4.6	4.55	1.32
Leu	6.55	7.6	7.9	7.1	6.9	8.3	6.95	7.55	1.27
Ile	3.6	4.4	5	3.75	4.2	4.25	4.1	4.1	1.39
Ser	4.95	5.2	5.15	4.9	5.25	6	4.85	5.2	1.24
Thr	2.7	3.5	3.9	3.45	3.55	4.3	3.8	3.55	1.59
Tyr	4	3.3	3.6	2.2	4.2	3.4	3.15	3.2	1.91
Phe	5.05	6.6	5.1	4.6	3.6	6.1	4.55	4.15	1.83
Trp	1.05		1.3	0.9	0.7		0.8	0.85	1.86
Pro	4.7	4.3	5.9	3.5	3.95	3.5	4	4.1	1.69
Met	1.4	1.4	1.55	1.6	0.7	1.15	0.95	0.7	2.29
Cys	1.45		1.65	1.9	1.7	1.1	1.7	1.35	1.73
Lys	3.45	7.2	6.25	7.3	4.5	6.7	7.2	6.35	2.12
His	2.2	2.3	2.75	2.6	2.05	2.7	2.35	2.45	1.34
Arg	11.6	8.8	8.1	6.9	9.95	6.7	9.55	9.6	1.73
Asx	11.7	11.7	11.8	12	10.3	12.4	11.5	11.7	1.2
Glx	18.9	16	18	15.8	21.5	15.9	17.1	17.6	1.36
Total	97.6	95	101.9	90.55	93.85	95.05	95.9	95.45	
Protein (%)	25	19.4	40	29	40	27	25	30	

For each amino acid the maximum and minimum contents are underlined with solid and dashed lines, respectively; protein = N × 6.25. References (A) Busson [13]; (B) Van Etten et al. [57]; (C) McCurdy et al. [34]; (D) Baudet and Mossé (unpublished results); (E) Baudet and Mossé [5].

The barley mutant *Hiproly* was found by Munck et al. [42] after a large screening of a world collection and then other induced mutant grains were obtained by Danish teams [32]. In a comparative analysis, Pomeranz et al. [49] have found a lysine content of 4.25 g per 16 g of nitrogen instead of the 3.3 gm of lysine observed in normal grains having nearly the same high protein content (18.6%) as the *Hiproly* mutant.

The sorghum mutant *hl* found by Singh and Axtell [52] has a lysine content of 3.3 g per 16 g of nitrogen with grains containing 15.7% of protein instead of the 2.05 g observed with normal grains containing 12.7% of protein.

In maize, Jimenez [30], Mossé [37], and Mossé et al. [40] have demonstrated that these biochemical mutations involve chiefly regulatory mechanisms. The synthesis of zein, the maize prolamin which is almost devoid of lysine, is repressed in favour of other proteins which contain lysine and which are synthesized in higher proportions than in normal grains. It is very striking that the particular biochemical control of such mutants seems often to affect also glucid biosynthesis and vice versa: Baudet et al. [2] have shown that the single recessive gene *amylose-extender* corresponds also to grains with higher levels of lysine. Many similar mutants have been subsequently analysed to show their better aminoacid composition from a nutritional point of view.

If these biochemical mutants are put aside, the literature is very short of available data on intraspecific or intravarietal variations of aminoacid composition. Many data do not allow any distinction between genetic factors and environmental or physiological conditions. Many of them also do not exhibit very significant changes in aminoacid composition, the determination of which is both complex and time consuming. But the main reason lies in the narrowness of the protein content ranges which have been investigated. One means of possibly solving the problem of limited data consists of working with a single variety grown in such various conditions that as large a range of protein content as possible can be obtained.

Relationship between seed protein content and aminoacid level

Such a broad range of seed protein content has been obtained by Coïc et al. [16] with the Rika variety of barley. From plants grown with very different levels of nitrogen, phosphorus, sulfur, potassium and/or water, ten barley seed samples were harvested with protein contents ranging from 9 to 17%. Aminoacid compositions were determined. If the amount of lysine (in g per 100 g of seed on a dry matter basis) is plotted against protein content [38], the occurrence of a linear relation is apparent with the exception of the only sample short of sulfur (Figure 3). In the same study, Mossé and Baudet [38] have analysed 29 different barley varieties for their grain aminoacid compositions and obtained the same linear relation between their lysine and protein contents. In other words, the relation which was observed appears as likely

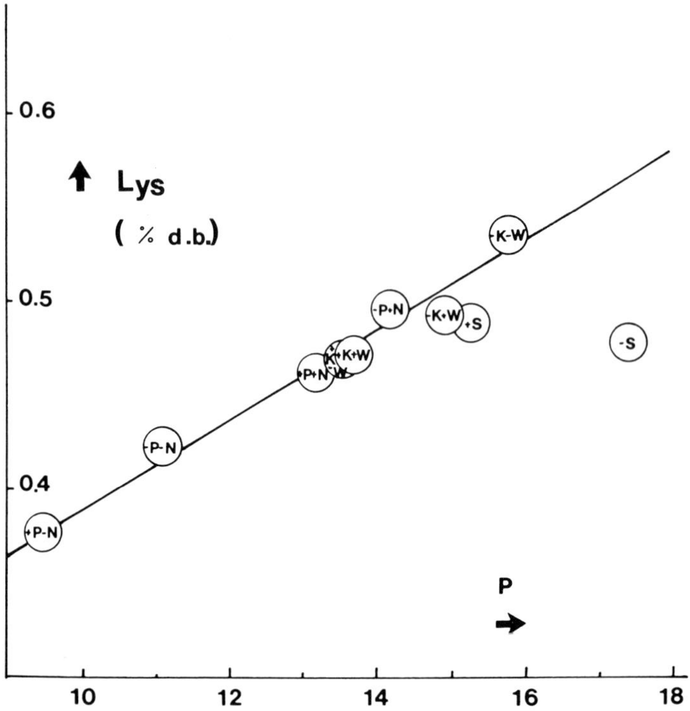

Figure 3. Seed lysine content (% db) as a function of seed protein content (N × 6.25; db) after data resulting from Coïc et al. [16] experiments on barley mineral nutrition; the barley grains were harvested from plants which were either normally supplied or starved in N, S, P, K or water. The sample out of correlation results from plants deprived of sulfur (-S).

to be generalized to any barley variety. The same kind of linear relation has been found, of course, for each protein aminoacid, but with obviously different parameters: for each aminoacid there is a corresponding straight line on the graph but with a slope and intercept which differ for each of the amino acids. For instance (Figure 4), complementary analyses on ten other barley samples have fully confirmed these results [39]. Nevertheless, some twenty barley mutant samples (*Hiproly* and *Bomi 1508*) were shown to obey the same kind of linear relationship, but to behave like a distinct species. All these results are in good agreement with those which can be calculated from the analyses made by Eppendorfer [22].

Similar studies have also been performed with oilseeds and legume seeds. For example, 70 samples of sunflower seeds ranging from 23 to 37% in protein (d.b.) have been analysed for their aminoacid composition by Baudet et al. [4]. Some of them correspond to the same line (or hybrid) harvested in different places and having different protein contents. Some of the others arise from distinct lines differing both by their genome and protein content.

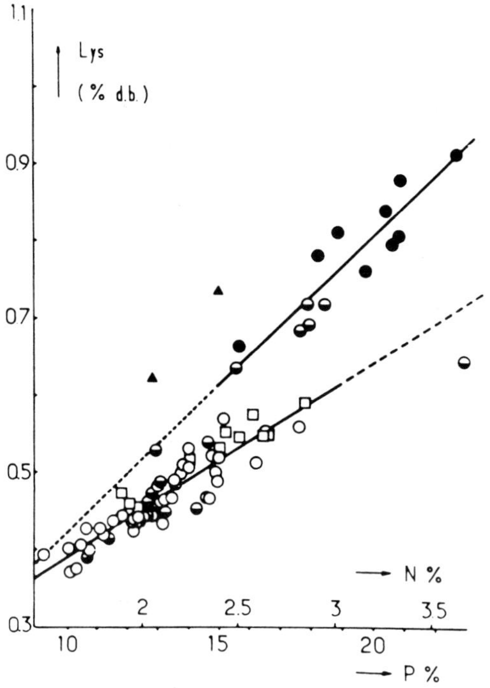

Figure 4. Seed lysine content (% db) as a function of seed nitrogen or protein (protein = N × 6.25 % db) content. Open figures correspond to normal husked (circles) or naked (squares) barley. Filled figures correspond to mutants hiproly (circles) and Risφ 1508 (triangles). Half filled circles correspond to hybrids between normal and hiproly barley (after Mossé and Baudet [39]).

Similar to the results with barley, a linear relationship is apparent (Figure 5). As it could be expected, the two straight lines corresponding to lysine have neither the same slope nor the same intercept when comparing barley and sunflower seeds; the same observation can be made with respect to each of the other protein aminoacids. For legumes, the aminoacid composition of 75 broadbean samples has also been determined [5]: their protein content varied from 23 to 38%. Figure 6 shows the straight lines obtained for the following aminoacids: Arg, Leu, Lys, Val, Phe, His, Cys, Trp and Met.

From a general point of view, if A_i is the seed content of one aminoacid i (in percentage of seed dry matter) and P its protein content (also expressed in percentage of seed dry matter), all the preceding results show the existence of the linear relation:

$$A_i = \alpha_i P + \beta_i$$

α_i and β_i being constant factors which characterize the aminoacid i in the species (α_i is the slope and β_i the intercept of the plotted straight line). It is

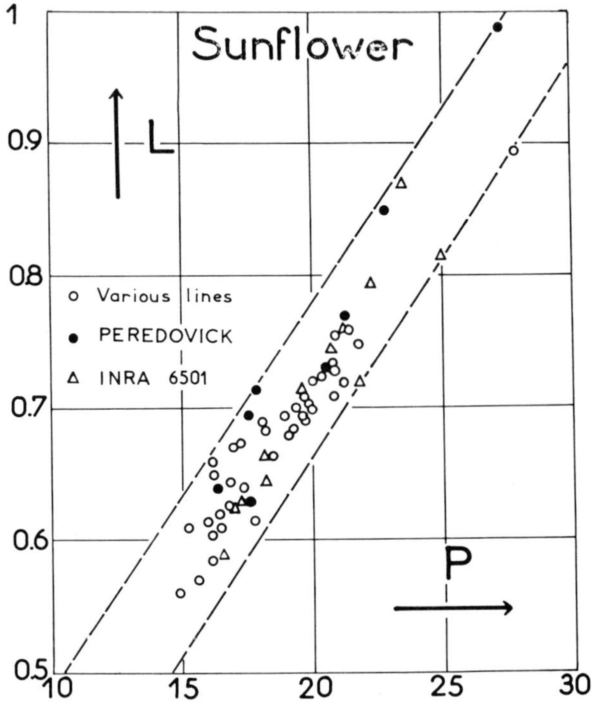

Figure 5. Seed lysine content (% db) as a function of seed protein content (N X 6.25 % db) in sunflower (after Baudet et al. [4]).

obvious that α_i cannot be negative: it is indeed impossible that any increase of the protein (P) gives rise to a decrease in the content of any aminoacid. But the factor β_i can be positive as well as negative (or equal to zero).

On the other hand, if a_i is the amount of one aminoacid i in the total seed protein (in percentage of crude protein, i.e. in g per 16 g of nitrogen), it can be written:

$$a_i = 100 \times A_i/P$$

The combination of the two preceding relations gives:

$$a_i = 100 \times (\alpha_i \mp \beta_i/P)$$

This last relation shows that variations of a_i in terms of P are represented by a segment of a hyperbola, the shape of which depends on both the sign and the value of β_i. Accordingly, as β_i is positive, nearly equal to zero or negative, a_i decreases, remains practically constant, or increases in terms of P. This is just the result observed in broadbean seed (Figure 7) for lysine, histidine and arginine, respectively.

In other words, in the latter example, the complementary proteins

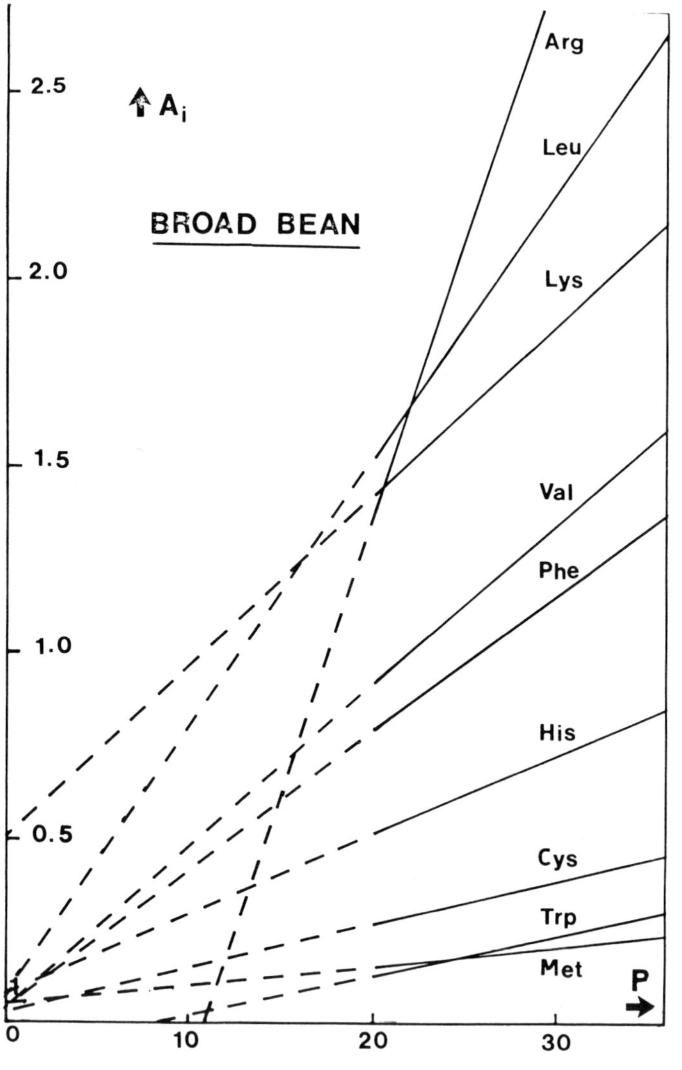

Figure 6. Variation in the contents of 9 different amino acids (A_i) of broadbean seeds (% db) as function of seed protein content (P); after Baudet and Mossé [5].

deposited in the high protein seeds are poorer in lysine, as high in histidine, and richer in arginine than the proteins accumulated in low protein seeds. This results from two kinds of data: (1) the proportion of legumin (one of the main legume seed storage proteins) is much higher in high protein seeds than in low protein ones [8, 61]; (2) legumin is poorer in lysine and richer in arginine compared to the other storage proteins.

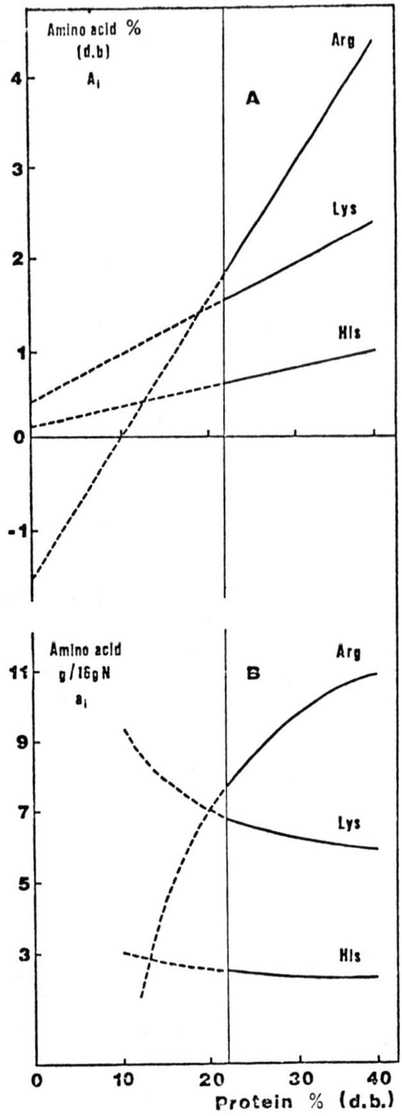

Figure 7. The effect of *Vicia faba* seed protein content (N × 6.25 g per 100 g of seed, on a dry matter basis) on basic amino acid levels expressed in two different ways: (A) Seed amino acid content (g per 100 g of seed, db) and (B) crude protein amino acid content (g per 100 g of protein); after Mossé and Baudet [39].

Several comments can be made about the variations of seed aminoacid composition in relation to protein content:

(1) It can be emphasized that the level of each proteic aminoacid in the seed of a botanical species is only dependent upon the seed protein content.

Therefore, every time that appropriate graphs (amino acid content vs. protein content) can be completely drawn for a species, only the knowledge of the seed crude protein content (i.e. of the nitrogen content) is needed to infer its aminoacid composition. The availability of such graphs would be particularly useful for the feed industry.

(2) The fact that each proteic aminoacid has its level which depends on a single factor, i.e. seed nitrogen content, leads to the conclusion that all of the proteic aminoacid contents depend one on the other one. Such correlations which have been mentioned by Dickson and Hackler [20], Pomeranz et al. [50], Hubbard et al. [29] and Lafiandra et al. [33] result only from the bound amino acid levels (aminoacid residues inside polypeptide chains), which precludes a consideration of any kind of independent variations of their seed content. The only exception to this rule may obviously concern the very rare proteic aminoacid of which a non negligible fraction remains free (non-protein) in the seed.

(3) The variation of the aminoacid composition of seeds in terms of their nitrogen content implies a change of the nitrogen to protein conversion factor. Although small, this change makes useless any search for a very high accuracy in the determination of this factor.

(4) The possibility for the β_i factor (or intercept) to be either positive or negative or equal to zero follows from the differences in velocity and rate of deposition of major seed storage proteins. Thus, in cereal grains, prolamins which are poor in lysine and rich in glutamine are more quickly accumulated than glutelin or other protein groups, as shown by Bishop [6] with barley. In the same way, in legume seeds, legumins which are poor in methionine, are deposited at a much higher rate than vicilins.

(5) Among cereal mutants, the straight lines representing seed aminoacid content in terms of protein content are shifted with a more or less significant slope modification. For instance, Figure 4 shows that the straight line for lysine has a much higher slope among mutant barley grains than among normal ones.

(6) The very general validity of these kinds of linear relationships suggests the occurrence of very accurate control mechanisms for the phenomena involved in seed storage protein accumulation.

(7) Among the ten barley samples represented in Figure 3, one point appears to fall outside of the linear relation. The grain sample involved resulted from plants drastically starved in sulfur. When deprived of sulfur, methionine and cysteine cannot be synthesized in sufficient amounts and a very different emergency control has to occur as suggested by the results of several authors like Finlayson et al. [26] in rape seed, Blagrove et al. [7] in lupin seed, Byers and Bolton [14] and Wrigley et al. [62] in wheat, and Millerd et al. [36] in pea seed.

Conclusions

It is now well established that the seed protein content of any species can vary widely. Now that modern analytical techniques allow for the rapid determination of protein content, significant progress can therefore be expected from plant breeding. Any change in seed nitrogen content, i.e. in crude protein content, results in a modification of the aminoacid composition. But seed aminoacid composition appears so strongly controlled by molecular mechanisms that only a very few mutants are shown to be able to escape.

Turning towards the future, two main approaches appear as useful, for improving the seed content of essential aminoacids, out of a systematic but blind screening which must also be efficient. These two approaches involve basic studies at the molecular level.

The first approach calls for the study of molecular mechanisms which control seed storage protein accumulation. Molecular biology studies have shown the possibility of investigating biosynthesis of seed storage proteins like the zein polypeptide chains in maize endosperm [10, 11, 12] or the main storage polypeptide chains of pea [9, 17, 18, 19]. The knowledge of these mechanisms is a condition for mastering them, even if it is a long and exacting (but exciting) task.

The second approach arises from works published very recently on seed storage proteins, especially on legume seed storage proteins. The latter are built up by association of elementary subunits (polypeptide chains) forming perfectly well-defined structures. Several authors (see [47] for review) have shown the possibility of obtaining in vitro molecular hybrids between subunits resulting from storage protein of dicotyledonous seeds belonging to very distant botanical families. These successful in vitro molecular hybridizations substantiate the existence of conformational homologies among many of these storage proteins. This conservatism was explained by the maximal packing hypothesis by Mossé and Pernollet [41, 47]. Genetic engineering can now allow in a few years the transfer of structural genes coding for storage proteins from one species to another.

In 1981, the gene coding for phaseolin – the major storage protein in *Phaseolus vulgaris* – has been transferred in *Helianthus* tumor by using as a vector a plant parasitic microorganism, *Agrobacterium tumefaciens,* which induces Crown-gall type tumors (Hall, personal communication). When enclosed in the sunflower genome, this gene is transcribed in messenger RNAs but not yet translated in polypeptide chains. This last difficult step will, in all probability, be overcome in a few years.

In other words, it would be possible, using a defined species, to choose storage protein molecules propitious for their high content in essential aminoacids and to substitute them, by genetic transfer, into another species having less favourably balanced proteins. The success of in vitro molecular hybridization related above, associated with the homologies in conformation and structure for several of these proteins, leads to the assumption that such substitutions will be viable and accordingly usable from a practical point of view.

References

1. Anonymous (1978) Interlaboratory comparison of amino acid analyses, 393–402. In: Seed Protein Improvement by Nuclear Techniques. Proc two Res Coord Meet Baden, 1977. IAEA Vienna
2. Baudet J, Cauderon A, Fauconneau G, Mossé J, Pion R (1968) Sur un troisième gène mutant (amylose-extender) qui accroît la teneur en lysine du grain de Maïs et sur son effet cumulatif avec le gène opaque 2. CR Acad Sci 266D:2260–2263
3. Baudet J, Cousin R, Mossé J (1977) Topographical study of weight and protein content of the Pea seed (*Pisum sativum*) following its localization on the plant, 316–323. In: Protein quality for leguminous crops (Commission of the European Communities – Coordination of agricultural research). EUR 5686 EN
4. Baudet J, Leclercq P, Mossé J (1971) Sur la richesse en lysine des graines de Tournesol en fonction de leur teneur en protéines. CR Acad Sci 273D:1112–1113
5. Baudet J, Mossé J (1980) Amino acid composition of different cultivars of broad beans (*Vicia faba*): comparison with other legume seeds, 67–82. In: Bond DA (ed), World Crops: Production, Utilization and Description, Vol 3. *Vicia faba*: feeding value, processing and viruses, Martinus Nijhoff, The Hague, Boston, London for the CEE 67–82
6. Bishop LR (1930) The composition and determination of the barley proteins. III. Fourth report on barley proteins. The proteins of barley during development and storage and in the mature grain. J Inst Brewing 36:336–349
7. Blagrove RJ, Gillespie JM, Randall PJ (1976) Effect of sulphur supply on the seed globulin composition of *Lupinus angustifolius.* Aust J Plant Physiol 3:173–184
8. Boulter D (1977) Quality problems in 'protein plants' with special attention paid to the proteins of legumes, 11–45. In: Protein quality from leguminous crops (Commission of the European Communities) Eur. 5686 EN
9. Boulter D (1983) Regulation of storage protein synthesis and deposition in developing legume seeds, 219–222 In: Daussant J, Mossé J, Vaughan JG (eds), Seed Proteins. Acad Press
10. Burr B, Burr FA (1976) Zein synthesis in maize endosperm by polyribosomes attached to protein bodies. Proc Nat Acad Sci USA 73:515–519
11. Burr B, Burr FA, Rubenstein I, Simon M (1978) Purification and translation of zein messenger RNA from maize endosperm protein bodies. Proc Nat Acad Sci USA 75:696–700
12. Burr FA, Burr B (1979) Molecular basis of zein synthesis in maize endosperm, 27–48. In: Rubenstein I, Phillips RL, Green CE, Gengenbach BG (eds), The Plant Seed Development, Preservation and Germination. Acad Press
13. Busson F (1965) Plantes alimentares de l'Ouest africain, étude botanique, biologique et chimique. Min Coop, Rech Sci & Techn et Armées
14. Byers M, Bolton J (1979) Effects of nitrogen and sulphur fertilizers on the yield, N and S content, and amino acid composition of the grain of spring wheat. J Sci Food Agric 30:251–263
15. Cherry JP (1977) Potential sources of peanut seed proteins and oil in the genus *Arachis.* J Agric Food Chem 25:186–193
16. Coïc Y, Fauconneau G, Pion R, Busson F, Lesaint C, Labonne F (1963) Influence de l'alimentation minérale sur la composition des protides des graines de Céréales (Blé et Orge). Ann Physiol Vég 5:281–292
17. Croy RRD, Gatehouse JA, Evans IM, Boulter D (1980a) Characterization of the storage protein subunits synthesized in vitro by polyribosomes and RNA from developing pea (*Pisum sativum* L.). I. Legumin. Planta 148:49–56

18. Croy RRD, Gatehouse JA, Evans IM, Boulter D (1980b) Characterization of the storage protein subunits synthesized in vitro by polyribosomes and RNA from developing pea (*Pisum sativum* L). II. Vicilin. Planta 148:57–63
19. Croy RRD, Gatehouse JA, Tyler ML, Boulter D (1980c) The purification and characterization of the third storage protein (convicilin) from the seeds of pea (*Pisum sativum* L). Biochem J 191:509–516
20. Dickson MH, Hackler LR (1975) Protein quantity and quality in high-yielding beans, 185–192. In: Nutritional Improvement of Food Legumes by Breeding (Proc Symp Roma 1972). J Wiley and Sons
21. Earle FR, Jones Q (1962) Analyses of seed samples from 113 plant families. Econ Bot 16:221–250
22. Eppendorfer WH (1975) Effects of fertilizers on quality and nutritional value of grain protein, 249–263. In: Proceeding of the 11th Coll of the Int Potash Institute. Bornholm (Denmark)
23. Eppendorfer WH (1977) Nutritive value of oat and rye grain protein as influenced by nitrogen and amino acid composition. J Sci Food Agric 28:152–156
24. Eppendorfer WH (1978) Effects of nitrogen, phosphorus and potassium on amino acid composition and on relationships between nitrogen and amino acids in wheat and oat grain. J Sci Food Agric 29:995–1001
25. F.A.O. (1970) Amino acid content of foods and biological data on proteins. Nutritional studies, no 24, Rome, 285 pp
26. Finlayson AJ, Bhatty RS, Christ CM (1969) Species and varietal differences in rapeseed. Can J Bot 47:679–685
27. Georgi B, Niemann EG, Brock RD, Axmann H (1979) Comparison of analytical techniques for seed protein and amino acid analysis, 311–341. In: Seed Protein Improvement of Cereals and Grain Legumes (Proc Symp Neuherberg, 1978). IAEA Vienna, vol II
28. Holt NW, Sosulski FW (1979) Aminoacid composition and protein quality of field peas. Can J Plant Sci 59:653–660
29. Hubbard JD, Pomeranz Y, Lai FS (1977) Note on protein contents and amino acid composition of dark hard and yellow hard kernels separated from red winter wheat. Cereal Chem 54:778–783
30. Jimenez JR (1966) Protein fractionation studies of high lysine corn, 74–79. In: Mertz ET, Nelson OE (eds), Proc of the High-Lysine Corn Conference. Corn Ind Res Found, Washington
31. Johnson VA, Mattern PJ, Kuhr SL (1979) Genetic improvement of wheat protein, 166–179. In: Seed Protein Improvement of Cereals and Grain Legumes (Proc Symp Neuherberg, 1978). IAEA Vienna, vol II
32. Kφie B, Doll H (1979) Protein and carbohydrate components in the Risφ high-lysine barley mutants, 205–215. In: Seed Protein Improvement of Cereals and Grain legumes (Proc Symp Neuherberg, 1978). IAEA Vienna, vol I
33. Lafiandra LD, Polignano GB, Golaprico G (1979) Protein content and amino acid composition of seed in varieties of *Vicia faba* L. Z. Planzenzücht 83:308–311
34. McCurdy SM, Scheier GE, Jacobson M (1978) Evaluation of protein quality of 5 varieties of lentils using *Tetrahymena pyriformis*. J Food Sci 43:694–697
35. Mertz ET, Bates LS, Nelson OE (1964) Mutant gene that changes protein composition and increases lysine content of maize endosperm. Science 145:279–280
36. Millerd A, Thomson JA, Randall PJ (1979) Heterogeneity of sulphur content in the storage proteins of pea cotyledons. Planta 146:463–466
37. Mossé J (1966) Alcohol-soluble proteins of cereals grains. Fed Proc 25:1663–1669
38. Mossé J, Baudet J (1969) Etude intervariétale de la qualité protéique des orges: taux d'azote, composition en acides aminés et richesse en lysine. Ann Physiol vég 11:51–66
39. Mossé J, Baudet J (1977) Relationship between amino acid composition and nitrogen contents of broad bean seed, 48–56. In: Protein Quality from Leguminous Crops (Commission of the European Communities Coordination of agricultural research) Eur 5686 EN

40. Mossé J, Baudet J, Landry J, Moureaux T (1966) Etude sur les protéines du Maïs. II. Comparaison entre les compositions en acides aminés et les proportions mutuelles des fractions protéiques de grains normaux et mutants. Ann Physiol vég 8:331–344
41. Mossé J, Pernollet JC (1982) Storage proteins of legume seeds. 111–193 In: Arora SK (ed), Chemistry and Biochemistry of Legumes. Edward Arnold, London
42. Munck L, Karlsson KE, Hagberg A, Eggum BO (1970) Gene for improved nutritional value in barley seed protein. Science 168:985–987
43. Nelson, OE (1969) Genetic modification of protein quality in plants. Adv Agron 21:171–194
44. Nelson OE (1979) Inheritance of amino acid content in cereals, 79–88. In: Seed Protein Improvement of Cereals and Grain Legumes (Proc Symp Neuherberg, 1978). IAEA Vienna, vol I
45. Nelson OE, Burr B (1973) Biochemical genetics of higher plants. Annu Rev Plant Physiol 24:493–518
46. Nelson OE, Mertz ET, Bates LS (1965) Second mutant gene affecting the amino acid pattern of maize endosperm proteins. Science 150:1469–1470
47. Pernollet JC, Mossé J (1983) Structure and location of legume and cereal seed storage proteins 155–191. In: Daussant J, Mossé J, Vaughan JG (eds), Seed Proteins. Acad Press
48. Pion R, Fauconneau G (1968) Les ressources mondiales actuelles et futures en protéines disponibles pour l'alimentation de l'homme, leurs possibilités d'utilisation, 3–59. In: Isotope Studies on the Nitrogen Chain (Proc Symp on the use of isotopes in studies of nitrogen in the soil plant animal system, 1967). IAEA, FAO Vienna
49. Pomeranz Y, Robbins GS, Gilbertson JT, Booth GD (1977) Effects of nitrogen fertilization on lysine, threonine, and methionine of hulled and hull-less barley cultivars. Cereal Chem 54:1034–1042
50. Pomeranz Y, Robbins GS, Smith RT, Craddock JC, Gilbertson JT, Moseman JG (1976) Protein content and amino acid composition of barleys from the world collection. Cereal Chem 53:497–504
51. Salamini F, di Fonzo N, Gentinetta E, Soave C (1979) A dominant mutation interfering with protein accumulation in maize seeds, 97–108. In: Seed Protein Improvement of Cereals and Grain Legumes (Proc Symp Neuherberg, 1978). IAEA Vienna, vol I
52. Singh R, Axtell JD (1973) High lysine mutant gene (hl) that improves protein quality and biological value of grain sorghum. Crop Sci 13:535–539
53. Singh U, Jambunathan R (1980) Evaluation of rapid methods for the estimation of protein in Chickpea (*Cicer arietinum* L). J Sci Food Agric 31:247–254
54. Sosulski FW, Holt NW (1980) Aminoacid composition and nitrogen-to-protein factors for grain legumes. Can J Plant Sci 60:1327–1331
55. Sosulski FW, Saswar G (1973) Amino acid composition of oilseed meals and protein isolates. J Inst Can Sci Technol Aliment 6:1–5
56. Tkachuk R (1969) Nitrogen to protein conversion factors for cereals and oilseed meals. Cereal Chem 46:419–424
57. Van Etten CH, Kwolek WF, Peters JE, Barclay AS (1967) Plant seeds as protein sources for food or feed. Evaluation based on amino acid composition of 379 species. J Agric Food Chem 15:1077–1089
58. Wolff G (1975) Relation between seed distribution in the pod, protein content, and thousand-kernel weight of *Pisum sativum*. Pisum Newsl 7, 60
59. Wolff G (1975) Distribution of seeds in the pods of *Pisum sativum* and some of its mutants. Pisum Newsl 7:61
60. World Health Organization (1973) Energy and protein requirements. WHO Tech Rep Series 552, FAO Nutrition Meetings Rep Series 52. Report of a joint FAO/WHO Ad Hoc Expert Committee
61. Wright DJ, Boulter D (1972) The characterization of vicilin during seed development in *Vicia faba* (L). Planta 105:60–65
62. Wrigley CW, du Cros DL, Archer MJ, Downie PG, Rowburgh CM (1980) The sulfur content of wheat endosperm proteins and its relevance to grain quality. Aust J Plant Physiol 7:755–766

Protein composition of grains of the *leguminosae*

DONALD BOULTER

Department of Botany, University of Durham, Science Site, Durham, U.K.

Key words: *Leguminosae*, peas, beans, storage proteins, structure and function

Abstract. The main seed proteins of Pisum and faba beans are the storage proteins synthesised in the seed and subsequently utilised to supply nitrogen and carbon for biosynthesis during germination and seedling development. There are two major families of storage proteins, legumin and vicilin. Legumins are hexameric proteins comprising of 6 M_r 60 000 subunits which, as isolated from mature seeds, show considerable charge and size heterogeneity. Vicilins are a less well-defined group of proteins consisting largely of M_r 50 000 subunits and non-equimolar amounts of lower molecular weight subunits. A third storage protein, convicilin, is also present in peas and beans. It is a vicilin-type protein with vicilin immunological determinants. Proteins homologous to legumin have been found in Arachis, Glycine, *Vigna unguiculata*, Lupinus, Cicer, Lens and Lathyrus. Equivalent vicilin-type proteins occur in *Phaseolus vulgaris*, Glycine, Arachis and *Vigna unguiculata*. The storage proteins have an unusual amino acid composition related to their biological role. They are multimeric and readily undergo association/dissociation reactions. They show a high degree of charge and size heterogeneity, partly caused by co and post-translational modifications and partly due to gene divergence.

Introduction

There are probably several thousand different proteins in the seeds of any one species of legume; different legumes contain many homologous proteins, although the precise extent of the homology is not fully certain. Legume proteins may be classified in a variety of ways [1], but the separation of seed proteins into the two categories, metabolic or storage proteins, is a simple and useful operative division.

In this review I shall discuss only the storage proteins, since these contribute most of the seed protein and are responsible for many of the characteristics of the seed protein fraction. It is not my intention to review exhaustively the legume seed storage proteins however (for general reviews see [8] and [20]), but to concentrate on those of peas (Pisum) and faba beans (Vicia) as models for storage proteins generally. However, grain legume seeds also contain various proteins such as enzyme inhibitors, lectins and other noxious or toxic proteins which are nutritionally or physiologically undesirable. The biological functions of these in the seeds is virtually unknown, although there is evidence in a few cases that they may help to protect against bacteria, fungi, insects and animals as part of the seed's relatively general defence mechanisms. Their physiological effects on humans are also little understood, although protease inhibitors are known to induce pancreatic hypertrophy in some animals, and lectins and other proteins can cause food poisoning and damage to the intestinal mucosa. These effects are normally lessened or off-set by appropriate preparation and cooking procedures.

Definition and function of storage proteins

Storage proteins are proteins produced in large quantity at one stage in the life cycle, to be used to supply nitrogen and carbon for biosynthesis at a subsequent stage; consequently, they must be stored, and in developing legume seeds this is accomplished, after their synthesis on the RER, by deposition in membrane-bound protein bodies [21]. Transport is via the ER and possibly the golgi apparatus, although precise details of the transport pathways are not as yet established [21].

Structure and subunit composition of storage proteins

In peas and beans, two major families of storage protein, legumin and vicilin, are found in the mature seeds. Legumins (M_r 360–400 000) are hexameric proteins comprised of six M_r^*. 60 000 subunits, each of which consists of a M_r^* 40 000 relatively acidic polypeptide, disulphide-linked to a M_r^* 20 000 relatively basic polypeptide [8]. These subunits, as isolated from mature seeds, show considerable charge and size heterogeneity [3, 4, 5, 9, 15, 17, 22]. In peas, the subunit pairs may be divided into three types [17]; the main type (M_r 53 000) corresponds to the conventional subunit pair of the Wright and Boulter, 1974 model [8] for Vicia legumin (the major legume subunits of Casey [2]) and the other two types, present in lesser amounts, have been called 'big', M_r 55–58 000, and 'small', M_r 35 000, legumin [17].

Vicia faba legumin contains as many as 10 disulphide-linked subunit pairs in the mol wt range 37 000–79 000 [18]. The use of two-dimensional gel techniques are extremely useful in this type of analysis and the following pairs were found using this technique [18]:

A (79 000) = (58 000) + (21 000)
B (73 000) = (54 000) + (21 200)
C (70 000) = (52 000) + (21 200)
D (67 000) = (50 000) + (21 200)
E (61 500) = (46 500) + (21 000)
F (60 000) = (43 500) + (21 000)
G (56 500) = (37 500) + (23 000)
H/I (53 000) = (35 000) + (23 000) and (35 000) + (21 600)
J (40 000–37 000; three bands) = (22 700–23 100) + (21 200–21 000).

* M_r of subunits is approximate as there is usually size heterogeneity; this notation is used to identify a type of subunit.

Further heterogeneity exists due to charge differences in subunits so that the number of potential species is very large indeed [18]. More size heterogeneity is found in the large relatively acidic subunits (23–58 000 M_r, pI range 4.6–6.1) than in the small subunits (21–23 000 M_r, pI range 8.2–8.5). The 37 000 and 21 000 subunits are present in by far the largest amounts. However, not all subunit pairs can be present in any one molecule since legumin is hexameric and many legumin molecules seem to be formed by a random association of subunit pairs although one subunit pair was shown to associate only with itself. Legumin of both peas and beans undergoes extensive co- and post-translational modification, including the removal of leader sequences and proteolytic cleavage, but not glycosylation. Thus, the observed heterogeneity in legumin as isolated is not due entirely to their being many divergent legumin gene copies (see later).

Vicilins ($M_r \sim 150\,000$) are a less well-defined group of proteins consisting largely of M_r 50 000 subunits and non-equimolar amounts of a range of lower molecular weight subunits, e.g. M_r 33 000, 29 000, 13 000, 12 500, some or all of which show size and charge heterogeneity [9, 23]. Only 50 000 and 47 000 M_r polypeptides are synthesised in vitro in cell free systems programmed with storage protein mRNA of peas and beans [5, 10]. Both polypeptide types have immunological determinants in common. The 47 000 M_r is not present in vicilin preparations isolated from mature seeds but can be detected in vitro from early developing seeds [5, 9, 23].

Vicilin is also extensively co- and post-translationally modified, including removal of leader sequences, proteolytic cleavage of some M_r 50 000 and 47 000 subunits to give smaller products and by glycosylation.

A third storage protein, convicilin, M_r 280 000, containing only M_r 71 000 subunits is also present in both peas and beans [6] which, apart from removal of leader sequences, does not appear to undergo co- or post-translational modification. It is a vicilin-type protein with vicilin immunological determinants and is not glycosylated.

Homology of the pea and bean proteins with storage proteins in other legumes

In 1976, Derbyshire et al. [8] suggested that legumin type storage proteins were distributed widely in the members of the Leguminosae and even in other dicotyledonous plants. Additional evidence obtained since then would tend to confirm this suggestion, at least for Arachis (arachin) [25, 26], Glycine (glycinin) [14], *Vigna unguiculata* [13], and Lupinus (α-conglutinin) [19]. Cicer, Lens and Lathyrus also have legumin-like storage proteins [11, 12]. Evidence for a legumin-type protein in Phaseolus is not so conclusive.

Vicilin-type proteins equivalent to those found in peas and beans have been shown to exist in *Phaseolus vulgaris* (phaseolin) [8], Glycine (conglycinin) [8], Arachis (conarachin) [8] and *Vigna unguiculata* [19].

The relationships suggested above are tentative, however, and more amino acid sequence data is needed. Many problems remain. For example, do storage proteins equivalent to γ-conglutinin of Lupinus (which doesn't easily fit into the vicilin/legumin families) exist in other legumes? What is the relationship of the seed proteins of *Canavalia ensiformis* which include urease and also canavalin, an inert precursor of α-D-mannosidase [16]? In several legumes, only very incomplete information on storage proteins exists [20].

Relationship of structure to function of storage proteins

The unusual amino acid composition of storage proteins relative to the average amino acid composition of metabolic proteins is related to their role in supplying nitrogen building blocks during germination and early development of the seedling. Their high amide and arginine contents ensure a relatively high nitrogen to carbon ratio of these proteins, an asset in a storage compound. Furthermore, these amino acids can be readily utilised by the enzymic machinery of the germinating seed and young seedling, mobilising them to provide a range of nucleic acid and protein building blocks [8].

Storage proteins are multimeric and readily undergo association/dissociation reactions. These properties allow their efficient packing in the protein body in an insoluble form so as not to contribute to the osmotic pressure within the protein body. Preliminary investigations (see for example [24] and Wright, these proceedings) suggest that it is the 'functional properties', e.g. structure and subunit composition, solubility characters, and hydrophobicity index, which have been preserved in evolution. This might be expected since these determine the packing and unpacking characteristics of the storage proteins. Since the three dimensional structure depends on the amino acid sequence, we can expect this also to be conserved and the limited data available, mostly unpublished, confirm this. Residue positions which are probably not important, however, can vary considerably. Thus the suggestion by some authors (see refs. in [8]) that specific types of storage proteins should be identified by their serological cross-reactivity is misleading.

The storage proteins show a high degree of charge and size heterogeneity of their constituent subunits. This is partly caused by co- and post-translational modifications [4, 5, 22] which are necessary (although as yet imperfectly understood) for the processes of transport and deposition. Some heterogeneity is, however, genic in origin since investigations of pea genomic DNA with cDNA probes indicates that about 4 legumin genes are present in single copies whereas the several vicilin genes occur in more copies but are not highly reiterated [7]. The significance as to why there are several different types of protein storage compound in any one legume is unclear.

Summary

The structure and functions of Pisum and faba bean seed storage proteins are described and their homology to storage proteins in other legumes briefly discussed, although insufficient amino acid sequence data is available to do this adequately.

Acknowledgements. It is a pleasure to acknowledge the help of Dr. J. Gatehouse in the preparation of this paper.

References

1. Boulter D, Derbyshire E (1978) The general properties, classification and distribution of plant proteins. In: Norton G (ed), Plant proteins. Butterworths, pp 3–24
2. Casey R (1979) Immunoaffinity chromatography as a means of purifying legumin from *Pisum* (Pea) seeds. Biochem J 177:509–520
3. Croy RRD, Derbyshire E, Krishna TG, Boulter D (1979) Legumin of *Pisum sativum* and *Vicia faba*. New Phytol 83:29–35
4. Croy RRD, Gatehouse JA, Evans IM, Boulter D (1980a) Characterisation of the storage protein subunits synthesised in vitro by polyribosomes and RNA from developing pea (*Pisum sativum* L). I. Legumin Planta 148:49–56
5. Croy RRD, Gatehouse JA, Evans IM, Boulter D (1980b) Characterisation of the storage protein subunits synthesised in vitro by polyribosomes and RNA from developing pea (*Pisum sativum* L). II. Vicilin Planta 148:57–63
6. Croy RRD, Gatehouse JA, Tyler ML, Boulter D (1980) The purification and characterisation of a third storage protein (convicilin) from the seeds of pea (*Pisum sativum* L). Biochem J 191:509–516
7. Croy RRD, Lycett GW, Gatehouse JA, Yarwood JN, Boulter D (1981) Cloning, characterisation and sequence analysis of cDNAs encoding *Pisum sativum* L. (Pea) storage protein precursors. Nature (submitted)
8. Derbyshire E, Wright DJ, Boulter D (1976) Legumin and vicilin storage proteins of legume seeds Phytochem 15:3–24
9. Gatehouse JA, Croy RRD, Morton H, Tyler M, Boulter D (1981) Characterisation and subunit structures of the vicilin storage proteins of pea (*Pisum sativum* L). Eur J Biochem (in press)
10. Higgins TJV, Spencer D (1981) Precursor forms of pea vicilin subunits. Plant Physiol 67:205–211
11. Jackson P, Boulter D, Thurman DA (1969) A comparison of some properties of vicilin and legumin isolated from seeds of *Pisum sativum*, *Vicia faba* and *Cicer* arietinum. New Phytol 68:25–33
12. Jackson P, Milton JM, Boulter D (1967) Fingerprint patterns of the globulin fraction obtained from various seeds of the Fabaceae. New Phytol 66:47–56
13. Khan MdRI, Gatehouse JA, Boulter D (1980) The seed proteins of Cowpea (*Vigna unguiculata* L Walp). J Exp Bot 31:1599–1611
14. Koshiyama I, Fukushima D (1976) A note on carbohydrates in the 11S globulin of soybean seeds. Cereal Chem 53:768–769
15. Krishna TG, Croy RRD, Boulter D (1979) Heterogeneity in subunit composition of the legumin of *Pisum sativum*. Phytochem 18:1879–1880
16. McPherson A, Smith SC (1980) Identification of canavalin as a proteolytically modified form of Jack bean α-D-mannosidase. Phytochem 19:957–959
17. Matta N, Gatehouse JA, Boulter D (1981a) Molecular and subunit heterogeneity of legumin of *Pisum sativum* L. (Garden Pea) – a multi-dimensional gel electrophoretic study. J Ex Bot (in press)
18. Matta N, Gatehouse JA, Boulter D (1981b) The structure of legumin of *Vicia faba* L. – a reappraisal. J Ex Bot 32:183–197
19. Mironenko AV, Troitskaya TM, Shurkhai SF, Domash VI (1978) Vicilin- and legumin-like proteins of yellow lupin seeds. Prikl Biokhim Mikrobiol 14:752–758
20. Mossé J, Pernollet JC (1981) Storage proteins of legume seeds (in press)
21. Pernollet JC (1978) Protein bodies of seeds: ultra-structure, biochemistry, biosynthesis and degradation. Phytochem 17:1473–1480

22. Spencer D, Higgins TJV (1980) The biosynthesis of legumin in maturing pea seeds. Biochem Int 1, 5:502–509
23. Thomson JA, Schroeder HE, Dudman WF (1978) Cotyledonary storage proteins in *Pisum sativum* I. Molecular heterogeneity. Aust J Plant Physiol 5:263–279
24. Wright DJ, Boulter D (1980) Differential scanning calorimetric study of meals and constituents of some food grain legumes. J Sci Food Agric 31:1231–1241
25. Yamada T, Aibara S, Morita Y (1979a) Dissociation–association behaviour of arachin between dimeric and monomeric forms. Agric Biol Chem 43:2549–2556
26. Yamada T, Aibara S, Morita Y (1979b) Isolation and some properties of arachin subunits. Agric Biol Chem 43:2563–2568

Legume seed improvement for protein content and quality

L.M. MONTI and S. GRILLO

Istituto di Agronomia Generale e Coltivazioni Erbacee, Cattedra di Miglioramento Genetico delle Piante Coltivate, Università degli Studi di Napoli, 80055 Portici, Italy

Key words: legume proteins, breeding, antinutritional factors, genetic traits

Abstract. The variability in seeds for protein content, the protein fraction and the aminoacid composition is reported for the main grain species. All of these features are under genetic control and the different possibilities of improving the species by genetic techniques are indicated, taking into account the genetic basis and the correlations of these traits with other characteristics of agronomic interest. The results of the breeding efforts performed by several authors show that progress can be obtained for improving both seed protein content and quality.

Several antinutritional factors have been found in the seeds of legume species. Whenever screening of the germplasm of a species has been carried out, a wide variability in the content of these substances was found. When the genetic base of these noxious substances was studied, it proved to be rather simple. A low content is inherited as a simple recessive factor; many lines have been isolated with a low content or none of these substances present.

Introduction

Although they are known to be an important source of protein, because of their high crude protein content, and high levels of some essential aminoacids, grain legumes have been neglected for a long time. In European and national research programmes, only recently has special attention been given to increasing grain legume yields, mainly for animal nutrition purposes. On the other hand, 70% of the total production of these crops in the world is used for human food. This Congress and others are an indication of the interest in Europe in the use of vegetable proteins as a source of food protein.

High possibilities are offered by the genetic techniques which are available for modifying the seed proteins of the legume species. Besides improving the seed yield, special care should be given to the biological quality of the legume products, in order to improve their use as human food. This paper reports the breeding efforts and the results obtained in selecting the main legume species for higher protein content and quality and for lowering or deleting antinutritional substances.

Protein content

Variation in protein content among different varieties is due to genetic differences and to the interactions of these differences with agronomic conditions

Table 1. Ranges of seed protein content (% d.m.) in legume species

Species	Protein content Min	Max
Cicer arietinum[a]	14.9	29.6
Glycine max[b, c]	34.8	40.5
Lens esculenta[c]	25.0	29.3
Lupinus albus[d, e]	17.0	38.7
Phaseolus vulgaris[a, f]	17.0	39.4
Pisum sativum[g]	15.5	39.7
Vicia faba[h]	22.0	37.0

[a] Mossé and Pernollet [49]; [b] Hove et al. [32]; [c] Salamini [61]; [d] Gardiner and Gorman [20]; [e] Green et al. [25]; [f] Leley et al. [41a]; [g] Slinkard [72]; [h] Sjödin et al. [69].

and physiologic conditions of the plants. Table 1 shows the range of the protein level reported in the literature for the main legume species. Minimum values of 15–17% have been observed and maximum values of 29–40%. In all of the species, a wide variability was found. Broad sense heritability estimates for this trait were 0.83 in white lupine [25], 0.4–0.8 in soybean [66, 36]; 0.75 in chickpea [62]; 0.30–0.64 in dry beans [41, 38]; 0.56–0.71 in peas [72], and 0.54 in *Vicia faba* [8]; these heritability estimates indicate good chances for improvement through selection. As far as the genetic control of the protein content is considered, partial dominance of low protein over high protein was found in dry beans [41], in *Vicia faba* [8] and peas [12], while the opposite was found in soybean [77, 76, 66].

Highly important in designing an effective breeding programme is the knowledge of the correlations between the character to be improved and other traits of agronomic interest. Table 2 reports the correlations found in the main legume species between the protein content and the grain yield.

A negative correlation has been found in all of the legume species between the yield and the seed protein content, but this correlation was significant only in a limited number of experiments. Negative but not significant correlations were found in *Vicia faba* by Bond (1977) [8], Abdalla and Gunzel (1979) [1], De Vries (1979) [14], Griffith and Lawes (1978) [27]; in *Pisum sativum* by Blixt (1978) [6] and Pandey and Gritton (1975) [53]; in *Glycine max* by Johnson et al. (1955) [36], and Hartwig (1969) [30] and in *Phaseolus vulgaris* by Conti and Zuffrano (1979) [11], Rutger (1968) [60] and Leleji et al. (1972) [41]. To the contrary, negative and significant correlations were found only by Slinkard (1980) [73] in peas, by Green et al. (1977) [25] in lupine, by Tandon et al. (1957) [74] and Kelly and Bliss (1975) [38] in dry beans. These differences can be due to the different materials and to the different environmental conditions, which both can have an influence on the strength of the correlations.

Among the yield components, the protein content is negatively correlated with the number of seeds per plant [68, 69, 70, 71, 63], while correlations

Table 2. Correlations between protein content of the seeds and other traits in five legume species

	Vicia faba	*Pisum sativum*	*Lupinus albus*	*Glycine max*	*Phaseolus vulgaris*
Protein content and seed yield	NS[e,f,g,h]	–[a] NS[i,j]	–[b]	NS[k,l]	–[c,d] NS[m,n,o]
Protein yield and seed yield		+[a]			+[o]
Protein content and 1000-seed weight	NS[p]	NS[q,r,s]	+[b]		
Protein content and n° of seeds per plant	–[p]			–[t]	
Protein content and seed position within the plant	+[u]	+[q]			
Protein content and lateness	+[v]	+[j]			

[a] Slinkard [73]; [b] Green et al. [25]; [c] Tandon et al. [74]; [d] Kelly and Bliss [38]; [e] Bond [8]; [f] Abdalla and Gunzel [1]; [g] Griffiths and Lawes [27]; [h] De Vries [14]; [i] Blixt [6]; [j] Pandey and Gritton [53]; [k] Johnson et al. [36]; [l] Hartwig [30]; [m] Conti [11]; [n] Rutger [60]; [o] Leleji et al. [41]; [p] Sjödin [71]; [q] Cousin [12]; [r] Slinkard [72]; [s] Gottshalk and Muller [23]; [t] Schuster and Posselt [63]; [u] Sjödin et al. [67]; [v] Picard [54].

Table 3. Expected response of the protein production in *Lupinus albus* by selecting for yield or for protein content; data from Green et al. [25]

	Grain yield (Kg/ha)	Protein content (%)	Protein production (Kg/ha)	Protein production Relative %
No selection	1705	38.4	655	100
Selection for yield	2013	37.4	754	115
Selection for protein content	1511	40.8	617	94

with 1000-seed weight were found to be not significant in field bean [68, 69, 70, 71] and in peas [12, 72, 23] and even positive in lupine [25] (Table 2). It seems possible to say that certainly a negative correlation between the protein content and the grain yield exists in the legume species as is also observed for cereals. However, this correlation seems to be small enough to allow selection of plants high in protein without reducing yield. Of course in this kind of breeding program, a selection for higher protein content should be performed only among lines yielding at least at the same level as the original populations; the production of proteins per hectar must be taken as a selection criterium.

The alternative possibility is to select for increased yield while trying to maintain the protein level constants. The choice between the two alternatives depends on the correlations found between the two traits in the experimental material. For example, Table 3 reports the effect on the total protein pro-

Table 4. Range of sulfur-aminoacid, tryptophan, and lysine content (g/100 g protein) in legume species

	Methionine	Cystine	Tryptophan	Lysine	References
Cicer arietinum	0.5–1.7	–	0.9	6.9–7.2	a,b,c
Glycine max	1.4–1.6	1.4–1.6	1.3–1.4	6.3–7.0	c,d
Lens esculenta	1.6	1.9	0.9–1.0	7.2–7.3	b,c
Lupinus albus	0.5–1.0	1.7–2.5	0.7–1.6	4.3–5.8	c,d,e,f,g
Phaseolus vulgaris	1 1–1.3	1.0	1.0	6.7–7.2	b,c h
Pisum sativum	0.6–1.0	0.8–1.7	0.7–1.9	6.9–8.2	c,e,i
Vicia faba	0 6–1.0	0.6–1.7	0.8–1.0	6.2–6.5	a,c,j,k
FAO standard	3.5		1.0	5.4	e

[a] Porceddu and Lafiandra [57]; [b] Porceddu et al. [56]; [c] Mossé and Pernollet [49]; [d] Williams [78]. [e] Evans and Bandemer [16]; [f] Ballester et al. [4]; [g] Duranti and Cerletti [15]; [h] Baldi and Salamini [3], [i] Holt and Sosulski [31]; [j] Evans and Boulter [17]; [k] Kaldy and Kasting [37]; [l] FAO/WHO [18].

duction in lupine when the selection is applied for grain yield or for protein level [25]. Because of the negative correlation between the two traits, a better response (115%) will be obtained when the selection is performed for the yield rather than for the protein content (94%). This is also confirmed by the correlations (Table 2) found between the protein yield and the seed yield in peas [73] and in dry beans [41].

We have to take into consideration that in a species where negative correlation between yield and protein content has been found, all the causes that determine a fertility reduction (pests and diseases, unfavourable weather and so on) can provoke an increase of the protein content. A further complication in the selection for higher protein level is the difficulty in sampling which has to be highly precise because of the influence on the protein content of the seed position in the plant. For example, in *Vicia faba* [67] and in pea [12], the higher the nodes at which the pods are, the lower the protein level in the seeds. As a correlation between protein content and lateness was also found in *Vicia faba* [54] and in peas [53], it would be interesting to study if a relation between the two findings might exist, on the basis of physiologic differences [56], due to competition between the vegetative and reproductive parts in the same plant and between plants with different biological cycles.

Besides the correlations, the influence of single genes on protein content has to be taken into account. In peas, the seeds with wrinkled cotyledons (a character which is due to the gene *r*) have about 3.4 % more protein than the smooth cotyledons (29.2% vs 25.8%). This is due to a pleiotropic effect of the *r* gene, which also causes a lower seed-weight. In fact, when the weight of protein per seed or yield of protein per hectar are considered, no difference

Table 5. The crude protein and sulphur-containing aminoacid contents (g/16 g N) in the seeds of 22 *Phaseolus* species (from Baldi and Salamini [3])

Species	Percent protein	Cystine	Methionine
Ph. aconitifolius	28.87	0.55	0.96
Ph. acutifolius	33.95	0.96	0.98
Ph. adenanthus	29.77	1.72	0.70
Ph. angularis	29.66	1.18	1.53
Ph. angustifolius	27.17	1.40	1.16
Ph. atropurpureus	32.18	1.51	1.14
Ph. bracteatus	31.44	1.35	1.23
Ph. calcaratus	27.52	0.99	1.42
Ph. caracalla	25.94	1.24	0.92
Ph. coccineus	25.32	1.19	1.08
Ph. dumosus	32.03	1.34	1.25
Ph. lathyroides	29.87	1.02	1.11
Ph. lunatus	23.81	1.02	1.40
Ph. mungo	32.39	0.76	1.59
Ph. panduratus	36.66	1.08	1.21
Ph. phyllanthus	22.57	2.28	1.56
Ph. radiatus	29.88	0.92	1.39
Ph. sp. 239.355	30.26	0.63	1.28
Ph. sp. 286.307	29.32	0.94	1.46
Ph. sp. 312.195	27.55	1.80	1.36
Ph. trilobatus	25.98	0.90	1.55
Ph. vulgaris	29.55	1.08	1.25
Ph. vulgaris (wild)	35.21	1.24	1.07

can be found between the two types of seeds, because the wrinkled seeds are lighter and give a lower grain yield.

To summarize, even in species in which a negative correlation has been found between yield and crude protein content, progress in breeding work can be obtained because enough variation is present in each species to allow the selection of plants with higher yields and a constant or even higher protein content. Unlike in the cereals, in these species it is still possible to select at the same time for both yield and protein level, because the physiologic competition between the glucide and protein production has probably not reached its limit.

Protein quality

In comparision with the FAO pattern of aminoacid requirements [18], the aminoacid composition of legume seeds is limiting in sulphur aminoacids (cystine and methionine) with tryptophan as the third limiting aminoacid. To the contrary, the legume seed proteins are particularly rich in lysine and in the other essential aminoacids. Table 4 reports the range which has been found for methionine, cystine, tryptophan and lysine in the main legume species by different authors. The variability does not seem very large, particularly in view of the differences in the analytical methods used and in the conditions under which the plants were grown.

Table 6. Correlation between protein content and aminoacid content (g/100 g protein) in the seed of *Vicia faba, Phaseolus vulgaris* and *Pisum sativum*

	Vicia faba	*Phaseolus vulgaris*	*Pisum sativum*
Protein content and arginine	+ [a,b,c,d]	+ [e]	+ [f]
Protein content and lysine	– [a,b,c,d]	– [e,g]	– [f]
Protein content and methionine	NS [h,i] – [a,b,c]	NS [e,g] – [i]	– [f]
Protein content and cystine	NS [h,i] – [a,c]	NS [e] – [j]	– [f]
Protein content and methionine + cystine.	– [a]	– [k]	

[a]Sjödin et al. [68]; [b]Griffiths and Lawes [26]; [c]Lafiandra et al. [40]; [d]Mossé and Boudet [50]; [e]Silbernagel [65], [f]Holt and Sosulsky [31]; [g]Tandon et al. [74]; [h]Frolich et al. [19]; [i]Rudolph et al. [59] [j]Adams [2]; [k]Conti and Zuffrano [11].

Many breeding programmes aim at getting lines with a better aminoacid composition, provided that a sufficient variation can be found, in the available germoplasm. For instance, in peas Holt and Sosulski [31] found a fairly good variability for methionine, tryptophan and cystine content. Obviously, the variability is greater when all the genus is considered. Analysing the inter-specific variability in *Phaseolus* (Table 5) Baldi and Salamini [3] found a high amount of sulphur aminoacids in one of the species (Ph. phyllanthus), which might be very useful in cross-breeding programmes.

However the progress which can be obtained in the direct selection of lines with better protein quality does not seem very easy: firstly, because of the above mentioned narrow variability for each aminoacid; secondly, because the variation in the content of an aminoacid is related to the levels of the others [40], since they are bound inside the polypeptide chains [49]; thirdly, because of the influence of the protein content (%) on the levels of each aminoacid (g/16 g N).

Whenever such studies have been performed (Table 6), negative and positive significant correlations between protein and lysine content and between protein and arginine content respectively, were always found; whereas negative but not always significant correlations were found between protein content and sulphur aminoacid content. This means that a possibility should exist for increasing the sulphur aminoacid content without decreasing the protein content. Nevertheless, when a selection for higher sulphur amino-acid content in *Vicia faba* was carried out, without considering the protein level, lines with low protein content and, as a consequence, with a lower aminoacid production were obtained [68, 69, 70, 71]. Table 7 reports the data from this experiment, with the expected responses for the methionine production in *Vicia faba*, when selection was applied for methionine content or for protein content. Considering a grain yield of 3,500 kg/ha, a higher methionine production (103%) was obtained after selection for an increased

Table 7. Expected response of the methionine production in *Vicia faba* by selecting for protein content or for methionine content (with a grain yield level of 3500 Kg/ha); data from Sjödin [71]

	Protein content (%)	Protein production (Kg/ha)	Methionine content (g/100 g prot)	Methionine production	
				(Kg/ha)	Relative %
No selection	29	1015	1.18	11.98	100
Selection for protein content	32	1120	1.10	12.32	103
Selection for methionine content	26	910	1.25	11.38	95

Table 8. Sulphur aminoacids and lysine content (mol %) in legumin (L) and vicilin (V) seed protein fractions in legume species

	Cicer arietinum[a]		*Glycine max*[b,c]		*Lupinus albus*[d]		*Phaseolus vulgaris*[e,f]		*Pisum sativum*[a,g]		*Vicia faba*[a]	
	L	V	L	V	L	V	L	V	L	V	L	V
Methionine	1.0	1.0	1.3	0.3	0.4	tr	1.5	0.7	0.5	0.2	0.6	0.4
Cystine	0.8	0.7	0.6	0.2	0.6	tr	0.6	0.2	0.8	0.2	0.8	0.2
Lysine	5.2	6.9	3.9	6.0	3.4	3.7	7.8	5.5	4.4	7.5	4.6	8.1

[a] Jackson, et al. [34]; [b] Gerritsen [22]; [c] Koshiyama [39]; [d] Duranti and Cerletti [15]; [e] Derbyshire et al. [13]; [f] Puszatai and Watt [58]; [g] Boulter and Derbyshire [9].

protein content (95%) than after selection for an increased methionine content, due to the negative correlations between methionine content and protein content. In any case, some lines with both high protein and sulphur aminoacid contents were obtained by the same author in *Vicia faba*, and in other species by other authors.

The negative correlation between protein content (%) and sulphur aminoacids (g/16 g N) is due to the fact that in the higher protein containing lines there is an increase of the storage proteins [10] which are characterized by a low sulphur aminoacid content. This was also confirmed in common beans, where the low protein lines contained about 20% albumin, while the high protein lines contained only 11% [79].

An improvement of the sulphur aminoacid content seems easier through an improved ratio either between the different globulin fractions or the different seed proteins. In all the legume species (Table 8), a higher sulphur aminoacid content is present in the legumin fraction than in the vicilin one, the latter having generally a larger amount of lysine. An increase of the legumin/vicilin ratio should be one of the goals to be reached, since the consequent decrease of the lysine would not have much importance, owing to the high amount of this aminoacid in the globulins. Such a programme has been started in *Vicia faba*. A variation for the above mentioned ratio was found and a genetic control of this ratio was ascertained [43]. The selection programme performed by Sjödin et al. (1981) [68, 69, 70, 71], was not

Table 9. Methionine, cystine and lysine (g per 100 g of amino acids in the hydrolyzate) in the albumin and globulin fractions of protein extracted from peas; from Bajaj et al. [5]

Fraction	Methionine	Cystine	Lysine
Albumin	1.1 ± 0.1	0.8 ± 0.1	10.6 ± 0.4
Globulin	0.5 ± 0.1	0.3 ± 0.04	7.9 ± 1.2

successful, since the plants selected with a better ratio showed a higher protein content and consequently a decreased content of sulphur aminoacids. It thus becomes a necessity to keep constant the protein level, when selecting for an increased legumin/vicilin ratio. In the cotyledon storage proteins of beans [79], the alkali-soluble fraction showed the highest concentration of methionine, so that breeding programmes in this species should aim to increase this fraction and a simplified procedure was proposed for this purpose. The other possibility is to increase the albumin/globulin ratio, considering the greater amount of both sulphur aminoacids and of lysine in the albumin fraction (Table 9). The respective amounts of protein fractions resulted under genetic control [51] and mutants with an altered ratio in favour of the albumins were found by Gottshalk et al. [24].

In conclusion, a possibility seems to exist to obtain, by genetic techniques, an improvement of protein quality in legume species by increasing their sulphur aminoacid content. As noted before, the main difficulties for a direct selection are due to the low variability which exists for each aminoacid, to the relationships among the aminoacids and to the negative correlations between the protein content and the sulphur aminoacid level. The breeding possibilities are better for increasing the protein or the globin fraction with a better aminoacid composition similar to methods already employed in cereal breeding. Particular attention in the selection work with legumes should, however, be given to the tryptophan content, which is the third limiting aminoacid in the legume species and which is also present only in limited amounts in the cereals.

Toxic and antinutritional substances

Several antinutritional substances have been found in the legume species. Their importance varies according to their amount, to their different effects (e.g. factors with toxic effects, factors which reduce protein digestibility) and to the possibility of their being reduced or eliminated in the different stages of food preparation and processing technologies. The main factors are listed in Table 10 for each legume species, taking into account the above mentioned considerations.

Favism is a hemolytic sex-linked disease which affects mainly children who are deficient in G6PD (glucose 6-phosphate dehydrogenase) when they consume faba beans. This inherited enzymatic deficiency affects over 100

Table 10. Main antinutritional factors present in legume seed

Faba bean	– favism inducing factors (Vicine, Convicine, L-DOPA) – trypsin inhibitors – tannins
Common bean	– antinutritional aminoacids – flatulent factors – lectins (phytohemagglutinins) – trypsin inhibitors – cyanogenetic factors
Soybean	– flatulent factors – lectins – trypsin inhibitors – goitrogenic glocosinolates
Lupin	– alkaloids
Common pea Chickpea	– trypsin inhibitors
Lentil	– cyanogenetic factor

million people, mainly in the Mediterranean area. The factors which are present in faba beans and which are supposed to be responsible for favism are vicine, convicine, divicine, isouramil and L-DOPA. Only recently, a quick and precise method was established for the screening of the lines with lower amounts of these substances [42]. To date, no genotypes lacking the favism-inducing-factors have been found, so that a necessity exists for performing large germoplasm screening and mutagenic experiments.

As to the trypsin inhibitors, significant amounts have been found mainly in soybeans and in dry beans. In soybean, there are different trypsin inhibitors (more than 6% of the total seed protein), the most important of which is known as the Kunitz trypsin inhibitor. After extensive screening, soybean lines lacking this factor were found and genetic studies have shown that the presence/absence of this inhibitor is controlled by a single gene, the absence being recessive [52]. In *Vicia faba*, a wide variability was found [44] for the content of trypsin inhibitors and the selection carried out by Sjödin et al. (1981) [68, 69, 70, 71] resulted effectively in the realization of lines with low levels of these substances.

The lectins (or hemagglutinins) are present in several legume species, mainly dry beans and soybeans. Studies have been performed in *Vicia faba* [68, 69, 70, 71] and a successful breeding programme was carried out for the selection of populations with low contents of these substances. The genetic basis is rather simple (one gene), with a low content of hemagglutinins as a recessive character [34a].

The flatulence factors, which are particularly important in common bean and soybean, are due to oligosaccharides, mainly raffinose and stachyose. By analysing several soybean cultivars and strains. Hymowitz and Collins [33] found a wide range in values of total sugar raffinose and stachyose contents and isolated some lines with a very low content of individual sugars. These offer good breeding opportunities.

As to the alkaloids, they are widely present in the seeds of Lupine species. V. Sengbush [64] was able to find alkaloid-free plants in *L. luteus* and in *L. angustifolius* after a screening of about 1.5 million plants; the absence of alkaloids was shown to be recessive.

The tannins are contained mainly in the testa of the seeds and are of particular significance in faba beans. Selection for a low content of total phenolics resulted in lines with larger seeds, because of the smaller proportion of the testa when compared to small seeds [68, 69, 70, 71]. From a breeding point of view, however, the selection for low-tannin lines is not a problem, because of the complete linkage between low tannin content and white flowers [55, 7, 68, 69, 70, 71].

In summary, whenever a screening on the germoplasm of a species has been carried out, a wide variability in the content of the various deleterious substances was found. For trypsin inhibitors, for lectins, for the flatulence factors, for the alkaloids and for the tannins, precise selection work has always given positive answers and results in isolation of lines with low contents of these factors. When the genetic base was studied, it proved to be always rather simple, with the low content inherited as a simple recessive factor. Through the analysis of the germoplasm and with peculiar selection techniques, there should not be any difficulties in obtaining varieties with improved nutritional quality. Also the mutagenesis can be successfully used, because the removal of these kinds of substances can be achieved by true gene recessive mutations and through all the different types of chromosome aberrations inactivating or eliminating the genes which are responsible.

From an evolution point of view, it must be said that the presence in the plants of toxic or bad-tasting substances has a precise meaning, since the animals are repelled from eating seeds, leaves and roots and this gives these plants an advantage under selection pressure. Examples of this positive function of the presence of antinutritional substances in plants have been reported in cowpea [21] and in broad bean [35] with regard to insect resistance, and in sorghum [29] with regard to bird resistance. A very careful analysis of all the consequences which might derive from the cultivation of new plant types lacking antinutritional substances should therefore be carried out.

Further perspectives for legume seed improvement

As we have shown, several studies are in progress with seed legume species and remarkable successes have been obtained already with traditional genetic techniques to make these plants more useful for human beings. By using special techniques such as *in vitro* culture and some chromosome induced mutants [48, 28], greater chances of genetic improvement of protein structure are offered. Also, mutagenesis should be stressed [46, 47], as an important tool in order to eliminate antinutritional factors. It is also true that by increasing our knowledge of nitrogen metabolism in these plants [56] and of the genetic basis of the variation in the seed protein fraction and components, increased genetic possibilities can be obtained for improving protein content and quality.

These studies are particularly advanced in peas, where five loci have been identified, determining five variant systems in the legumin and vicilin fractions of storage proteins [75], and where some of the loci responsible for vicilin and convicilin fractions have been localized on chromosomes 7 and 2, as reported by Boulter (these proceedings).

References

1. Abdalla MMF, Gunzel G (1979) Protein content and electrophoresis of seed proteins of certain *Vicia faba L.* stocks and their assumed ancestors. Z Pflanzenzchtg 83: 148–154
2. Adams MW (1973) On the quest for quality in field bean. PAG Symp on 'nutrition improvement of food legumes by breeding', Rome (Italy), July 3–5, 1972, 142–149
3. Baldi G, Salamini F (1973) Variability of essential aminoacid content in seeds of 22 *Phaseolus* species. Theoret Appl Genetics 43:75–78
4. Ballester D, Yanez E, Garcia R, Erazo S, Lopez F, Haardt E, Cornejo S, Lopez A, Pokniak J, Chichester CO (1980) Chemical composition, nutritive value, and toxicological evaluation of 2 species of sweet lupine (*Lupinus albus* and *Lupinus luteus*). J Agric Food Chem 28:402–405
5. Bajaj S, Mickelsen O, Lillevik HA, Baker LR, Bergen WG (1971) Prediction of protein efficiency ratio of peas from their albumin content. Crop Science 11: 813–815
6. Blixt S (1978) Natural and induced variability for seed protein in temperate legumes. FAO/IAEA int symp seed protein improvement in cereals and grain legumes, Neuherberg, Sept 4–8, 1978
7. Bond DA (1976) In vitro digestibility of the testa in tannin-free field beans (*Vicia faba L*) J Agric Sci, Cambridge 86:561–66
8. Bond DA, (1977) Breeding for zero-tannin and protein yield in field beans (*Vicia faba L*) EEC Seminar on 'protein quality from leguminous crops', Dijon (France), November 3–5, 1976
9. Boulter D, Derbyshire E (1971) Taxonomic aspects of structure of legume protein 285–30. In: Harbone J, Boulter D, Turner BL (eds), Chemotaxonomy of the Leguminosae. London: Academic Press
10. Boulter D (1977) Quality problems in 'protein plants' with special attention paid to the proteins of legumes. EEC seminar on protein quality from leguminose crop, Dijon (France) November 3–5, 1976
11. Conti L, Zuffrano M (1979) Variabilità intra interspecifica della quantità e qualità delle proteine del seme di fagiolo. CNR Meeting on 'prospettive delle proteaginose in Italia', Perugia (Italy), May 30–June 2, 1979
12. Cousin R (1979) Breeding for protein and yield in pea. CNR meeting on 'prospettive delle proteaginose in Italia', Perugia (Italy), May 30-June2, 1979
13. Derbyschire E, Wright DJ, Boulter D (1976) Legumin and vicilin, storage protein of legume seeds. Phytochem 15:3–24
14. De Vries APH (1979) In search of characters to be used for indirect selection on grain and protein yield in *Vicia faba L.* EEC seminar on 'some current research on *Vicia faba* in Western Europe', Bari (Italy), April 1978, 324–341
15. Duranti M, Cerletti P (1979) Aminoacid composition of seed proteins of *Lupinus albus*. J Agric Food Chem 27:977–78

16. Evans RJ, Bandemer SL (1967) Nutritive value of legume seed protein. J Agric Food Chem 15:439–443
17. Evans IM, Boulter D (1980) Crude protein and sulphur aminoacid content of some commercial varieties of peas and beans. J Sci Food Agric 31:238–242
18. FAO/WHO (1973) Expert group of energy and protein requirements. World Health Org Techn Rep Ser No 522
19. Frolich WG, Pallmer WC, Christ W (1974) Variation of the contents of protein and methionine and cystine in *Vicia faba L.* Z Pflanzenzuchtg 72:160–169
20. Gardiner ML, Garman RC (1965) Austr J Exp Agric Anim Husbandry 5:410–16
21. Gatehouse AMR, Gatehouse JA, Dobie P, Kilminster AM, Boulter D (1979) Biochemical basis of insect resistance in *Vigna unguiculata*. J Food Agric 30:948–58
22. Gerristen T (1956) Aminoacid composition of the globulin from *Lupinus angustifolius* and *Lupinus luteus*. Biochim Biophys Acta 22:269
23. Gottshalk W, Muller HP (1974) Quantitative and qualitative investigation on the seed proteins of mutants and recombinants of *Pisum sativum*. Theoret Appl Genetics 45:7–20
24. Gottschalk W, Muller HP, Wolff G (1976) Further investigations on the genetic control of seed production in *Pisum* mutants. Proc Symp, Vienna, IAEA, 157–177
25. Green AG, Oram RN, Read BJ (1977) Genetic variation for seed yield, protein content, oil content and seed weight in *Lupinus albus*. Aust J Agric Res 28:785–93
26. Griffiths DW, Lawes DA (1976) Genetic variation in the protein content and composition of field beans (*Vicia faba*) in relation to improvement by plant breeding. EEC seminar on 'protein quality from leguminous crop', Dijon (France), November 3–5, 361
27. Griffiths DJ, Lawes DA (1978) Variation in the crude protein content of field beans (*Vicia faba*) in relation to the possible improvement of the protein content in the crop Euphytica 27:487–495
28. Grillo S, Saccardo F, (1983) Induced duplications in chromosome segments in peas and their significance for seed protein studies. EEC seminar on 'perspectives for peas and lupins as protein crop', Sant' Agnello-Sorrento (Italy), October 19–22:72–78
29. Harris HB (1969) Bird resistance in sorghum. Proc 24th Am Corn and Sorghum Res Conf, Chicago, pp 113–122
30. Hartwig EE (1969) Breeding soybeans for high protein content and quality. New approaches to breeding for improving plant protein. IAEA, Vienna 1969, STI/PUB, 212:67–70
31. Holt NW, Sosulski FW (1979) Aminoacid composition and protein quality of field peas. Can J Plant Sci 59:653–660
32. Hove EL, King S, Hill GD (1978) Composition, protein quality, and toxin of seed of the grain legumes, *Glycine mas, Lupinus* spp, *Phaseolus* spp, *Pisum sativum* and *Vicia faba* NZJ. Journal of Agric Res 21:457–462
33. Hymowitz T, Collins FI (1974) Variability of sugar content in seed of *Glycine max (L)* Merril and *G. soya* Sield and Zucc. Agronomy J 66:239–240
34. Jackson P, Boulter D, Thurman DA (1969) A comparision of some properties of vicilin and legumin isolated from seed of *Pisum sativum, Vicia faba* and *Cicer arietinum*. New Phytol 2:25–33
34a. Jaffé WG, Brucher O (1972) Toxicidad y especifidad de diferentes fitohemagglutininas de frijoles (Phaseolus vulgaris) Arch Latinoamer Nutr 22:267
35. Janzen DH, Juster HB (1976) Insecticidal action of the phytohemagglutinin in black beans on a bruchid beetle. Science 192:795–96
36. Johnson HW, Robinson HF, Comstock RE (1955) Genotypic and phenotypic correlations in soybean and their implications in selections. Agronomy J 47:477–83
37. Kaldy MS, Kasting R (1974) Amino acid composition and protein quality of eight faba beans cultivars. Can J Plant Sci 54:869–871
38. Kelly JD, Bliss FA (1975) Hereditability estimates of percentage seed protein and available methionine and correlations with yield in dry beans. Crop Science 15: 753–757
39. Koshiyama I (1968) Chemical and physical properties of a 7s protein in Soybean globulins Cereal Chem 45:394–404

40. Lafiandra D, Polignano GB, Colaprico G (1979) Protein content and aminoacid composition of seed in varieties of *Vicia faba L*. Z Pflanzenzuchtg 83:308–314
41. Leleji OI, Dickson MH, Crowder LV, Bourke JB (1972) Inheritance of crude protein percentage and its correlations with seed yield in Beans, *Phaseolus vulgaris L*. Crop Science 12:168–71
42. Marquardt RR (1981) Favism. International conference on faba beans Cairo (Egypt), March 7–11
43. Martensson P. (1980) Variation in legumin vicilin ratio between and within cultivars of *Vicia faba L* var *minor*. Proc EEC seminar, Cambridge, June 27–29, 1979
44. Mc Nab JB (1977) Trypsin inhibitors in field beans (*Vicia faba L.*) Seminar on 'protein quality from leguminous crop', Dijon (France), November 3–5, 1976
45. Monti ML (1977) Le leguminose da granella per l'alimentazione animale: il miglioramento genetico della *Vicia faba L* e sua importanza per l'Italia. Rivista di Agronomia 4:229–235
46. Monti ML (1968) Mutations in peas by diethylsulphate and x-rays. Mutation Research 5:189–191
47. Monti ML, Saccardo F (1979) The gamete irradiation technique in pea genetics and breeding. Monografia IV Genetica Agraria 141–152
48. Monti ML (1983) Natural and induced variability in peas for protein production. EEC seminar on 'Perspectives for peas and lupins as protein crop', Sant' Agnello-Sorrento (Italy), October, 19–22, 1981, pp 23–39
49. Mosse J, Pernollet JC (in press). Storage proteins of legume seed
50. Mosse J, Bourdet J (1977) Relationship between aminoacid composition and nitrogen contents of broad bean seed. EEC seminar on 'protein quality from leguminous crop', Dijon (France), November 3–5, pp 48–57
51. Muller HP (1973) Content in globulins, albumins and essential aminoacids in seeds of pea mutants and recombinants. Pisum News Letter 5:37–38
52. Orf JH, Hymowitz T (1979) Inheritance of the absence of the Kunitz Trypsin inhibitor in seed protein of Soybeans. Crop Science 99:107–109
53. Pandey S, Gritton ET (1975) Genetypic and phenotypic variances and correlations in peas. Crop Science 15:353–356
54. Picard J (1979) Some reflections on problems and prospects in *Vicia faba* breeding. Proc EEC seminar, Bari (Italy), April 27–29, 1978
55. Picard J (1976) Apercu sur l'hérédité du caractère absence de tannins dans les graines de fèverole (*Vicia faba L.*) Ann Amél Plantes 26:101–06
56. Porceddu E, Scarascia Mugnozza G, Monti LM (1979) Genetica e miglioramento genetico della produzione proteica delle leguminose da granella. CNR Meeting on 'prospettive delle proteaginose in Italia', Perugia (Italy), May 30-June 2, 1979
57. Porceddu E, Lafiandra D (1980) Il contributo della genetica al miglioramento delle proteine vegetali. Agric Ital 109:15–47
58. Puszatai A, Watt WA (1970) Glycoprotein II, The isolation and characterization of a major antigenic and non-haemagglutinating glycoprotein from *Phaseolous vulgaris*. Biochim Biophys Acta 207–413
59. Rudolph A, Hanelt P, Lehmann Ch, Muntz K, Scholz F (1975) Protein-screening am Weizen-und Ackerbohnen-Sortiment in Gatersleben. Nahrung 19:739–807
60. Rutger JN (1968) Variation in protein content and its relation to other characters in beans, *Phaseolus vulgaris* I. Agon Abs 20
61. Salamini F (1975) Variabilità genetica per quantità e qualità delle proteine della leguminose con particolare referimento al genere *Phaseolus*. Atti Simp Inter Alim ed il problema proteico in America Latina, Roma (Italy), November 6–9, 1973
62. Sandhu SS, Keim WF, Hodges HF, Nyquiest WF (1968) Inheritance of protein and sulphur content in chick peas, *Cicer arietinum* I. Agronomy Abst 20
63. Schuster W, Posselt V (1977) Protein content and protein quality of some soybean varieties on different locations. EEC seminar on 'protein quality from leguminous crop', Dijon (France), November 3–5, 1976, pp 324–338
64. Sengbush R von (1948) Susslupinen und Ollupinen. Landw J 91:720–880

65. Silbernagel MJ (1971) Bean protein improvement work by USDA-bean and pea investigations. Rep 10th Dry Beans Res Conf, Davis (Calif), August 12–13, 1970
66. Singh L. (1969) Maternal and cytoplasmatic effects on protein content of soybean seed. Diss Abst (Int) 30:2992-B
67. Sjödin J, Martensson P, Magyarosi T (1977) The variation in crude protein content and seed weight within field bean plants (*Vicia faba L.*) analyzed with the half-seed technique. J Swedish Seed An 87:33–42
68. Sjödin J, Martensson P, Magyarosi T (1981) Selection for improved quality in field beans (*Vicia faba L*) Z Pflanzenzuchtg 86:221–230
69. Sjödin J, Martensson P., Magyarosi T (1981) Selection for increased protein quantity in field bean (*Vicia faba L*) Pflanzenzuchtg 86:210–220
70. Sjödin J, Martensson P, Magyarosi T (1981) Selection for antinutritional substances in field bean (*Vicia faba L.*) Z Pflanzenzuchtg 86:231–247
71. Sjödin J (1981) Protein quantity and quality in *Vicia faba.* International conference on faba beans, Cairo (Egypt), March 7–11
72. Slinkard AE (1976) Breeding for protein in peas in Canada. Eucarpia Pea Meeting, Norwich (England), April 6–7
73. Slinkard AE (1980) Breeding field peas for increased yield and protein. In: Production, Utilization and Marketing of Field Peas. Final report.
74. Tandon OB, Bressani R, Scrimshaw NS, Le Beau (1957) Nutritive value of beans, nutrients in Central American beans.J Agr Food Chem 5:137–42
75. Thomson JA, Schroeder HE (1978) Cotilodonary storage proteins in *Pisum sativum.* II Hereditary variation in components of the legumin and vicilin fractions. Aust J Plant Physiol 5:281–294
76. Weber CR (1950) Inheritance and interrelation of some agronomic and chemical characters in an interspecific cross in Soybean *Glycine max G ussuriensis.* Iowa State Agr Exp Sta Bull 374
77. Williams LF (1948) Inheritance in a species cross in Soybean. Genetics 33:131–32
78. Williams W (1979) Studies on development of Lupine for oil and protein. Euphytica 28:481–88
79. Yu MA, Bliss FA (1978) Seed proteins of common bean. Crop Science 18:431–437

Legume seed protein extraction, processing, and end product characteristics

J. GUEGUEN

Ministère de l'Agriculture, Institut National de la Recherche Agronomique, Centre Agro-Alimentaire, Laboratoire de Biochimie des Aliments, 44072, Nantes-Cédex, France

Key words: Legume seed, concentrate, isolate, extraction, processing, antinutritional factors, functional properties

Abstract. Because of the difficulties in growing soybean in many parts of the world, other leguminosae crops (fababean, pea, lentil, lupine, bean chickpea, cow pea, etc.) are now being studied as new protein sources. They generally have a high protein content and a satisfactory amino acid composition. The studies which have led to the development of industrial flow sheets for protein extraction have mainly been carried out on pea (*Pisum sativum*), fababean (*Vicia faba*), and lupine (*Lupinus albus*). The processes generally used are (1) pin-milling plus air-classification which when applied to starch-rich legume seeds (pea, fababean), results in concentrates (defined as having protein contents of 60–75%), and (2) wet processes which produce isolates (defined as having protein contents of 90% to 95%). By air-classification, concentrates having 68% and 65% protein can be obtained, respectively, from fababean (31% protein) and pea (21% protein). Isolates, prepared by extraction of the flour proteins with alkaline solution followed by acid precipitation, have a protein content (N × 6.25) generally between 90% and 96% and a protein recovery yield varying between 60% and 65%. As a rule, isolates resulting from ultrafiltered extracts have a higher protein content. From the nutritional evaluation of these two types of products, concentrates and isolates, it appears that wet processes are more efficient for eliminating antinutritional factors. α-galactosides and glycosides are present in isolates only in traces. As for trypsin inhibitors and haemagglutinins, only one third of the activities in the flour remained in the isolates (in dry processes, the residual levels of these antinutritional factors were higher). From the studies of their functional properties, it appears that isolates and concentrates from sources such as fababean and peas, produced by the new processes described, are to some extent complementary or equivalent in their functional properties to those of the soybean, particularly for emulsifying and foaming purposes. These observations should encourage the development of these new processes.

Introduction

The leguminosae constitute one of the most important botanical families from an economic point of view, providing, especially with the Papilionaceae subfamily, a wide range of food and feed sources mainly as seeds and fodder. Among the different tribes of *Viciae*, *Phaseoleae*, *Lupinae*, *Glycineae*, *Diocleae*, and *Trifoliae* can be found most of the world's protein-rich vegetables such as, for example, fababean or peas (*Viciae*), mungbean or limabean (*Phaseoleae*), lupines (*Lupinae*), soybean and peanut (*Glycinae*), and clover or alfalfa (*Trifoliae*).

In fact, legume seeds constitute the basic protein source in the diets of many developing countries. In the developed countries, they are used mainly as an important protein-rich fodder in intensive animal production but they are also of importance (especially in the case of soybean products) as meat substitutes or functional agents in the food industry. Because of the difficulties in growing soybean in many parts of the world, other leguminosae have been, and are being, studied as vegetable protein sources.

In Europe, many studies have been carried out on fababean *Vicia faba* L., pea, *Pisum sativum* L. and lupine, *Lupinus.* Plant protein-rich products, e.g. protein concentrates or protein isolates, are needed to decrease the consumption of meat by the production of meat-like foods and of other new protein-rich foods. Moreover, the demand from the food industries for these products as technological or functional agents is now increasing.

In developing countries, as a part of their policy of increasing food production and availability, legume plants for human consumption have also been developed as a supply of good protein and an adequate complement to cereals. Such species as cowpea (*Vigna unguiculata*) and chickpea (*Cicer arietinum*) are being studied. But for economical and social reasons, these countries cannot as a rule use manufactured products like isolates or concentrates. They preferably must develop cheaper processes to improve the quality of the protein-rich raw materials. This partly explains why more research has been done on European legume seeds and less on tropical legume crops for isolate and concentrate processing.

For these reasons and because fababeans, peas and lupines have begun to be grown to quite a large extent in Europe, only the work done on these seeds, and work concerning protein extraction processes, biochemical composition and functional properties of their end products will be considered here. Results from studies which have been carried out on soybeans will be reported only as reference data. In order to make clear how the different protein processing procedures have been developed with respect to the various raw materials, it is appropriate to first discuss their biochemical compositions.

Biochemical composition of some legume seeds

The common legume seeds can be divided into two main categories: (a) those containing lipids and (b) those containing starch as energy storage components (Table 1). The former are mostly found in Lupinae and Glycinae subfamilies to which the lupine and soybean species belong, while in the second group belong Viciae and Phaseolae which include the pea, fababean, lentil and cowpea. The chickpea can be considered as an exception because of its relatively high contents of both lipids and starch.

From these composition data, it appears that in the starch-rich seeds most of the non-protein material is insoluble in usual solvents while in soybean

Table 1. Biochemical compositions of some legume seeds[a]

Subfamily	Cultivar	Protein (N X 6.25)	Crude fat	Crude fiber	Starch
	Fababean[b] (*Vicia Faba minor*)	34.5	1.8	2.4	47.0
	Pea[c] (*Pisum Sativum*)	24.4[d]	2.0	N.D.	48.6
Viciae	Chick pea[e] (*Cicer arietinum*)	19.5	7.4	3.3	50.0
	Lentil[e] (*Lens culinaris*)	23.9	1.1	3.8	52.8
	Kidney bean[e] (*Phaseolus vulgaris*)	24.0	1.7	4.5	40.3
Phaseolae	Cow pea[f,g] (*Vigna unguiculata*)	28.0	1.9	5.0	40.6
Lupinae	Lupin[f,h] (*Lupinus albus*)	39.5	12.5	13.4	–
Glycinae	Soybean[f,i] (*Glycine max*)	42.8[d]	19.6	5.5[d]	–

[a]Results are expressed in % on a dry matter basis.
[b]From Cerning et al. [7].
[c]From Colonna et al. [9].
[d]Calculated from the original data.
[e]From Sosulki and Youngs [44].
[f]Whole seeds.
[g]From Longe [26].
[h]From Ballester et al. [3].
[i]From Smith and Circle [43].

most of them are soluble. The nitrogen-free extract content is close to 30% for soybean. For these reasons, the principles used for protein concentrate or isolate processing are different according to the biochemical composition of the seeds. In soybean technology, the different constituents of the seed are extractable by solvents, the oil part by hexane (producing the meal), the soluble carbohydrates by acid or alcoholic solutions (leading to the concentrates), and the protein by an alkali solution (producing the isolates). In starch-rich legume seeds, the use of wet processes can only lead to the production of protein isolates. The low amount of soluble carbohydrate does not allow for the use of wet processes for protein concentrates preparation. Due to the biochemical composition of the seeds, it can be established that the fababean concentrates produced by wet processes can contain only about 38% protein whereas the protein level can be as high as 68% when soybean flour is used (Figure 1).

Lupines, as non-starch legume seeds, are closer to the soybean. The biochemical composition of the white lupine seed, for example, is near that of wild soybean cultivars with a low oil content (5.4%) and high protein (49.5%) and crude fiber (10.3%) contents [43]. However, the lipid contents of

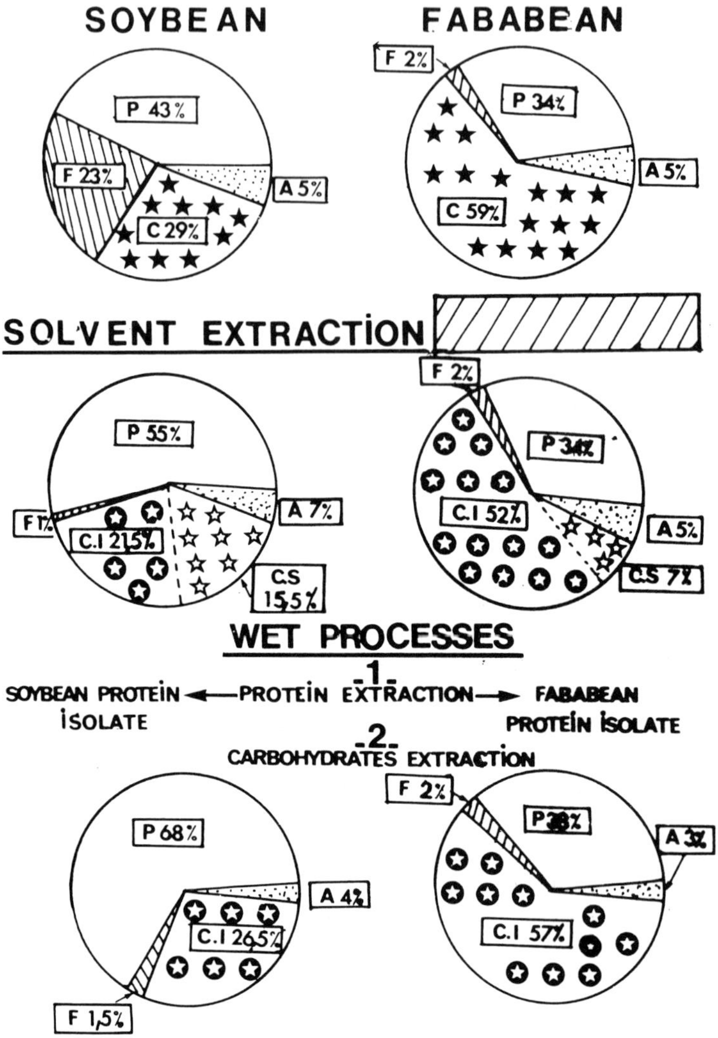

P: Protein

F: Fat

C: Carbohydrate { S Soluble / I Insoluble }

A: Ash

Figure 1. Evaluation of the soybean technology used for fababean.

today's lupine varieties are not high enough to allow, from an economic point of view, for industrial oil extraction processing. This means therefore that increasing the lipid content of lupine by breeding would make it possible to process them as soybeans.

Storage protein composition

Storage proteins in most legume seeds are mainly composed of globulins (Table 2). This protein fraction is higher in soybean and lupine (~ 90%) than in fababean and pea (60% and 66%, respectively). The globulins in the common legume species are composed of two major proteins characterized by their sedimentation coefficients which in most cases approach 7S and 11S. (In pea and fababean, the 7S and 11S globulins are called vicilin and legumin, respectively.) The ratio between these two globulins differs from one species to another. In soybean and lupine, the 7S-like protein is found in a larger proportion than the 11S-like protein; the 7S:11S ratios being about 1.6:1 [46] and 1.3:1 [15], respectively. On the other hand, legumin is the major protein in fababeans and peas; the vicilin:legumin ratio (7S:11S ratio) is close to 1:2. This probably explains some of the differences between the physiochemical properties of the protein, as a whole, of the different legume seeds.

From the results obtained by Gatehouse et al. [20] and Martensson [27] it appears that the vicilin:legumin ratio can also vary to a large extent from one cultivar to another. It varies between 1:1.3 and 1:4.2 for peas [20] and between 1:1.6 and 1:3.7 for fababean [20, 27]. But the influence of these variations on the protein extraction conditions is limited because, since their isoelectric points are found in the same pH range, the pH solubility characteristics of these proteins are very similar. For example, the isoelectric points are pH 4.8 for legumin and pH 5.5 for vicilin in *Pisum sativum* [12]. Moreover, these proteins are reported to be both located in the protein bodies [51, 29]. Thus, every mechanical treatment such as milling or screening does not alter the ratio of the two globulins.

Other proteins such as trypsin inhibitors and hemagglutinins must also be

Table 2. Composition of proteins in legumes

	Albumins[a] (g/100 g protein)	Globulins[a] (g/100 g protein)	Glutelins[a] (g/100 g protein)	Vicilin:Legumin ratio
Fababean	20	60	15	1:1.6[b]–1:3.7[c]
Pea	21	66	12	1:1.3[b]–1:4.2[b]
Soybean	10	90	0	1.6:1[d]
Phaseolus	15	75	10	
Lupinus	10–20	80–90	0	1.3:1[e]

[a]From Boulter [5].
[b]From Gatehouse et al [20].
[c]From Martensson [27].
[d]From Thank and Shibasaki [46].
[e]From Duranti et al. [15].

considered in the protein extraction technology because of their antinutritional activities.

Antinutritional components

Legume seeds are known to contain several antinutritional components such as trypsin inhibitors, hemagglutinins, flatulence factors, and glucosides. As shown in Table 3, soybean defatted meal is richer than most of the usual legume seeds in trypsin inhibitors and hemagglutinating factors, containing about seven times more trypsin inhibitor activity and ten times more hemagglutinin activity than fababeans. These levels of antinutritional factors are generally decreased for soybean by heating the defatted flour. Under these conditions the values are close to those obtained for other legume seeds.

Table 3. Trypsin inhibitor and hemagglutinating activities of some legume flours

	Trypsin inhibitor activity (T.U.I./mg dry matter)	Hemagglutinating[a] activity (units/mg dry matter)
Fababean[b] (*Vicia Faba*)	5.6–11.8	25–100
Pea[b] (*Pisum sativum*)	4.5–9.3	100–400
Kidney bean[c] (*Phaseolus Vulgaris*)	13.2	–
Cow pea[b] (*Vigna unguiculata*)	21.1	< 0.05
Lupine[b] (*Lupinus albus*)	< 1	0.1
Soybean (*glycine max*)[d]		
Raw flour	70	
Defatted flour	85	1600–3200
Industrial meal (toasted)	5.5–0.63	25–200

[a]All hemagglutinating activity data are expressed in hemagglutinating activity units as defined by Valdebouze et al. [48] which differs from the units as defined by Liener.
[b]From Valdebouze et al. [48].
[c]From Rouanet and Besançon [42].
[d]From Valdebouze (unpublished results).

As for flatulence factors (raffinose, stachyose and verbascose), *Viciae* subfamily species (especially fababean and pea), do not seem to contain high amounts (Table 4). Fababean, pea, and lupine contain smaller amounts (2.50%–4.5%) of total oligosaccharides than chickpea, cowpea, and soybean (7.0%–7.85%). Moreover, in both chickpea and soybean, stachyose is present in a high proportion. In contrast, the *Vicia* species like fababeans contain the glucosides, vicine (0.7%) and convicine (0.2%) [35, 39] which could lead to a disease known as favism in some sensitive individuals.

This brief summary of the biochemical composition of the main legume species should be useful for understanding how the different protein

Table 4. Oligosaccharide contents (% of dry matter) of some legume flours

	Fababean[a] (*Vicia Faba*)	Pea[a] (*Pisum Sativum*)	Chickpea[b] (*Cicer arietinum*)	Cowpea[c] (*Vigna ungiculata*)	Lupine[a] (*Lupinus Albus*)	Soybean[d] (*glycine max*) defatted flour
Sucrose	2.30	2.40	1.70	1.6	4.3	7.80
Raffinose	0.20	0.45	0.80	0.7	1.0	1.25
Stachyose	0.50	1.35	6.45	2.7	2.8	6.30
Verbascose	1.80	2.75	0.60	3.6	traces	–
Raff. + Stach. + Verb.	2.50	4.55	7.85	7.0	3.8	7.55

[a]Quemener and Mercier [39]. [b] Aman [1]. [c]Longe [26]. [d]Cegla and Bell [6].

extraction processes are to be applied to each material. With these data in mind, the yields and end-product characteristics of the different legume seeds are compared below for dry processes and for wet processes.

Protein extraction processes

Dry processes

Pin-milling of starch-rich legume seeds like peas and fababeans leads to flours containing two populations of particles differentiated by both size and density. This phenomenon was exploited, using air classification techniques to separate the light or fine fraction (the protein concentrate) from the heavy or coarse fraction (the starch fraction) [52].

A flow sheet of this process, mainly developed at the Prairie Regional Laboratory (Saskatoon, Canada), is shown in Figure 2. Whole or dehulled seeds are ground very finely and the flour is classified in a spiral air stream, separating out the starch. The fractionation can be repeated one or several times through the pin-milled and air-classifier. After one run, some protein bodies still adhere to the surface of the starch granules; agglomerates

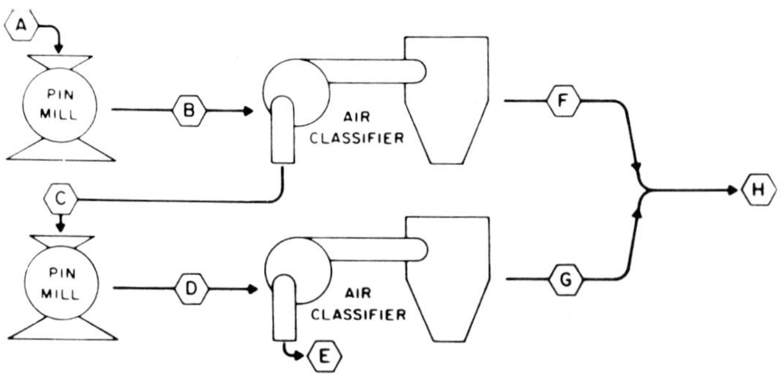

FABABEAN

A	100 LBS dehulled beans	(31% protein)	E	59 LBS	bean starch	(4% protein)
B	100 LBS bean flour	(31% protein)	F	22 LBS	bean protein	(70% protein)
C	78 LBS bean starch	(20% protein)	G	19 LBS	bean protein	(65% protein)
D	78 LBS bean starch	(20% protein)	H	41 LBS	bean protein	(68% protein)

PEA

A	100 LBS whole peas	(21% protein)	E	65 LBS pea starch	(20% protein)
B	100 LBS pea flour	(21% protein)	F	25 LBS pea protein	(60% protein)
C	75 LBS pea starch	(8% protein)	G	10 LBS pea protein	(46% protein)
D	75 LBS pea starch	(8% protein)	H	35 LBS pea protein	(56% protein)

Figure 2. Schematic flow sheet for air-classification of fababean and pea; from Vose et al. [52].

remaining in the initial starch fraction consist of starch granules embedded in a protein matrix [41]. By repeated pin-milling and air-classification, the starch fraction can be purified. But even after four runs, the protein content is still 2.16% and, because of this, its recovery yield decreases from 65% to 40% for peas [41] (Figure 3). The protein content of the light fractions P II, P III, and P IV, resulting successively from the repeated regrinding and air-classification of the starch fraction, are respectively 27.56%, 8.14% and 4.64%. It appears from these Canadian studies that a process based on more than two runs is unnecessary.

With a two-run process, the protein fraction is obtained with a yield of 41% and 35% and a protein content of 68% and 56% for fababean and pea, respectively. The biochemical composition of each fraction resulting from the air-classification of pea flour is given in Table 5. The protein fraction is enriched in lipids and ash. Comparative experiments on whole and dehulled pea and fababean seeds have shown that dehulling has only a small effect on the crude fiber content of the high protein fraction. In fact the milled hulls are recovered in the heavy starch fraction (Table 5). Concerning the starch fraction, the α-amylase susceptibility test indicates that pin-milling and air-classification do not cause damage to the starch granule. However, by scanning electron microscopy, a slight exfoliation on the granule surface can be observed [9].

Whereas the results are very similar for fababeans and peas, experiments carried out on wrinkled peas have shown air-classification to have a very low efficiency for separating the protein from the starch for this raw material. That could be due, on the one hand, to the higher lipid content of the initial

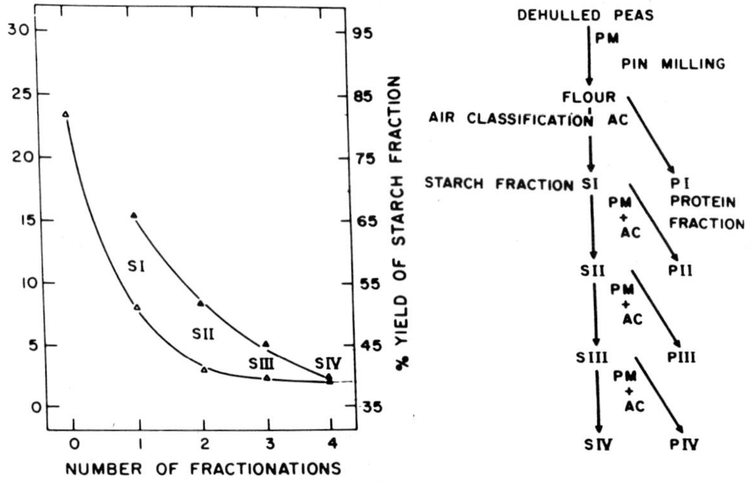

Figure 3. Purification of the starch fraction by repeated pin-milling and air-classification: (a) schematic diagramme, (b) yield (▲) and protein content (△) of starch fractions; from Reichert and Youngs [41].

Table 5. Composition (% dry matter) of light and heavy fractions from air classified field pea flour[a]

	Dehulled peas					Whole peas		
	Flour	Protein P_I	Fraction P_{II}	Starch S_I	Fraction S_{II}	Flour	P_I	S_I
Protein (N × 6.25)	29.1	60.5	40.0	14.5	4.8	25.7	63.4	13.3
Starch	46.6	7.8	27.6	63.0	78.0	43.7	7.6	60.3
Oil (a) neutral	0.9	1.8	1.4	0.46	0.17	0.9	2.2	0.48
Oil (b) polar	1.3	2.3	2.6	0.96	0.37	1.5	2.7	0.77
Ash	2.9	5.6	4.4	1.6	0.8	2.7	5.8	1.6
Sugars	N.D.	11.2	9.7	5.3	3.8	N.D.	N.D.	N.D.
Crude fiber	1.4	2.2	2.6	1.1	0.5	6.8	2.1	8.4
Phytic acid	0.74	1.9	N.D.	0.2	N.D.	N.D.	N.D.	N.D.

[a]From Vose et al. [52]; N.D. = not determined.

flour (4.5%), preventing a good separation of the particles in the air stream, or, on the other hand, to the broad distribution of the starch granule sizes [9]. The granule size of the isolated broadbean and smooth pea starches is narrowly distributed between 26.5 μ and 23.75 μ, while the starch granules of wrinkled peas have diameters ranging from 6 μ to 80 μ [9].

By experimenting on smooth green pea flour, a white protein concentrate and a green starch fraction can be obtained. But regrinding and refining the starch fraction leads to a decreasing chlorophyll content in the starch fraction and an increased green coloration in the light protein fraction [41] (Figure 4). Sosulski and Youngs (1979) [44] have studied the efficiency of pin-milling and air-classification on eight legume seeds and have compared the yields observed to those obtained on non-starchy materials, lupine and soybean flours. The light material which represents 22.5% to 29% of the original flours has a protein content of 47.7% to 66.6% (except for chickpea) and contains most of the flour lipids and ash while the coarse fraction contains 51% to 68% starch and only 12.2% to 15.6% protein. Thus in most of these species the separation of the protein from the starch material is very efficient (Table 6). In chickpea, as in wrinkled pea, the high level of lipids significantly decreased the efficiency of the fractionation [41]. Air-classification of the pin-milled defatted soybean meal and lupine flour appears to be of no benefit for increasing the protein level in the light fraction. A significant portion of the protein is left in the heavy material (Table 6). Such a process for

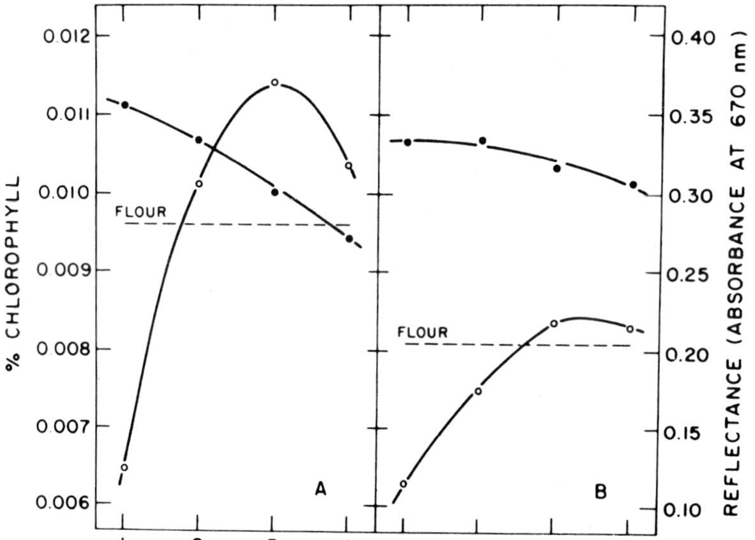

Figure 4. Partitioning of chlorophyll in air-classified flour fractions; • starch fractions, ○ protein fractions (from Reichert and Youngs [41]).

Table 6. Composition (% dry matter) of legume flours, protein fractions (P.F.) and starch fractions (S.F.)[a]

	Crude protein			Starch			Crude fat		
	Flour	P.F.	S.F.	Flour	P.F.	S.F.	Flour	P.F.	S.F.
Soybean (*Glycine max*)	52.5	54.2	50.5	4.7	–	–	0.6	–	–
Lupine (*Lupinus augustifolius*)	41.4	43.3	30.0	3.5	–	–	7.6	–	–
Chickpea (*Cicer arietinum*)	19.5	28.9	15.3	50.0	30.3	57.3	7.4	10.8	5.8
Pea bean (*Phaseolus Vulgaris*)	24.7	52.4	15.2	38.4	1.4	51.6	1.7	3.3	1.0
Northerm bean (*Phaseolus Vulgaris*)	24.0	53.5	15.6	40.3	1.4	51.5	1.7	3.9	1.3
Fababean (*Vicia Faba minor*)	29.8	66.6	14.4	42.4	1.4	57.5	1.3	2.6	0.8
Field pea (*Pisum sativum arvense*)	25.3	61.3	14.5	45.9	2.1	60.0	1.1	2.6	0.7
Lima bean (*Phaseolus lunatus*)	23.0	47.7	13.9	45.5	0.0	61.1	0.9	2.0	0.7
Mung bean (*Vigna radiata*)	26.5	60.4	12.3	50.0	6.1	67.7	0.9	2.2	0.6
Lentil (*Lens culinaris*)	23.9	57.9	12.2	52.8	7.5	68.2	1.1	2.4	0.7

[a]From Sosulski and Youngs [44].

producing fababean protein concentrate, on an industrial scale, is being operated in France by the 'Grandes Minoteries à Fèves de France'. The equipment used is a pin-mill Alpine Contraplex A 400 CW and an air-classifier Alpine 'Mikroplex' A 400 MS. The main difference in comparison to the Canadian flow-sheet is that, in the French operation, for economic reasons only one run is used. Under these conditions, 1000 to 1200 kg fababean are treated per hour, leading to the production of both starch and protein fractions which contain, respectively, 18% to 24% and 50% to 70% protein. The recovery yield, in weight of the concentrates, is generally between 15% and 25%. Today, a protein concentrate (65% protein), a protein enriched flour (50% protein) and a starch fraction which still contains 22%–24% protein are commercially produced; the former two mainly for human consumption with or without texturization and the latter for animal feeding [50].

Wet processes

In order to prepare protein isolates, the most widely used process is the one patented by Anson and Pader [2]. After an alkaline solubilization of the proteins, the insoluble material is removed by centrifugation. By adding hydrochloric acid to the supernatant, the protein is precipitated isoelectrically, thereby producing the isolate.

This process, tested on fababean some time ago by Flink and Christiansen [18], is very well suited to processing the seeds of legumes because most of them contain a legumin-like and a vicilin-like globulin protein. Although the ratio of these two globulins varies from one species to another, because of the similar pH solubility characteristics of the two proteins, the pH protein solubilization curves obtained for the flours of the common legumes are very similar. By studying the nitrogen extraction profile on nine legume flours, Fan and Sosulski [17] have shown that their proteins are easily extractable in an alkaline solution and are less solubilized in the acid pH range around pH 4. The great difference between the maximum and the minimum values for dispersible nitrogen over the pH range explains the high protein recovery yields which vary between 64.3% (lima bean flour protein) and 87.6% (mungbean flour protein) (Table 7). Only soybean, lupine and fababean isolates contain greater than 90% protein ($N \times 6.25$) but the mungbean and field pea isolates also have high levels of protein, ~88% and ~87%, respectively. These general studies, conducted on a laboratory scale, also demonstrate why interest exists in fababean, lupine and pea as protein sources for protein isolates.

At the I.N.R.A. Center of Nantes (France), fababean and pea flours are studied on a laboratory scale as well as on the pilot-plant level. The experiments carried out on a laboratory scale show that dehulled fababean flour has to be used preferably in order to avoid dark coloration of the isolates, due to the polyphenolic components of the hulls. In contrast, whether pea is dehulled or not does not interfere with the protein isolate

Table 7. Yield of protein isolate from legume seeds[a]

	Flour	Protein isolate			
			Yields		
	Protein (%)	Nitrogen (%)	g/100 g flour	% Recovery of total protein	Color
Soybean (*Glycine max*)	43.5	15.0	36.6	78.9	Cream
Lupine (*Lupinus albus*)	44.7	15.2	30.8	65.6	White
Fababean (*Vicia Faba minor*)	32.8	14.9	28.2	80.2	Tan
Pea bean (*Phaseolus vulgaris*)	31.4	13.1	28.6	74.6	White
Mung bean (*Vigna radiata*)	27.1	14.1	26.9	87.6	Yellow
Field pea (*Pisum sativum arvense*)	24.9	14.0	22.7	79.8	Cream
Lima bean (*Phaseolus lunatus*)	21.8	12.5	17.9	64.3	White
Lentil (*Lens culinaris*)	21.7	13.3	19.0	72.8	Cream
Chickpea (*Cicer arietinum*)	21.0	13.6	18.5	74.6	Cream

[a]From Fan and Sosulski [17]; values given are % dry matter except as otherwise specified.

characteristics. But, to a large extent, if the non-dehulled seeds are used, the composition of the starchy by-product recovered after protein extraction is modified with an increase of its cellulose and hemicellulose contents. In most cases the use of dehulled flour is preferred because of the rather low protein content of the hulls (3%–7%) which, if included, decreases the protein content of the raw starting material.

An average particle size of the flour between 100 μ and 150 μ is chosen as being the most suitable. Studies on fababean flours having different particle size distributions in the range of 80 μ to 810 μ show that an increase of the particle size does not significantly change the optimum solubilization pH or the recovery yield but increases the proportion of extract remaining in the starchy by-product. When coarse flour (particle size greater than 210 μ) is used instead of a fine flour, the extract remaining in the pellet reaches 25% of the whole extract instead of the 10% observed when a laboratory centrifuge was used.

The influence of solubilizing agents has also been systematically studied. Potassium or sodium hydroxide gave about the same solubilization yields, whereas calcium hydroxide solubilized less than 10% of the protein. In the acid pH range, at pH 1.5, about 85% of the flour proteins were extracted by most of the commonly used acids (HCl, H_3PO_4, lactic) but not by sulfuric acid which only gave a solubilization rate of around 40% [21].

The precipitation of the alkali protein extract can be performed by hydrochloric acid as well as by sulfuric acid. Under these conditions, around 75% of the nitrogen in the extract of both peas and fababeans is precipitated at pH 4.5, representing about 80% to 90% of the extracted protein nitrogen. The quantities of sodium hydroxide and sulfuric acid used are related to the protein content of the flour. The normality of the extraction solution generally used can be varied between 0.015 N and 0.020 N. As for the precipitating agent, 0.25 to 0.45 eq. H^+ are used per kg of flour. Based on these laboratory studies, pilot-plant experiments were carried out by a continuous process using equipment with a flow rate which could be varied between 700 l/hour and 1200 l/hour. The flow sheet is shown in Figure 5. Loads of 10 kg dehulled pea or fababean industrially ground flours are mixed discontinuously in tap water (100 l) in a mixing container. The slurry is pumped into a continuously stirred batch where the protein solubilization takes place for a duration of 20 min. Before acid precipitation, the protein extract is first separated from the starchy by-product by means of a Decanter centrifuge and then clarified in a solids-ejecting centrifuge. The acidification is then performed by injecting sulfuric acid into the extract tubing, to obtain the right pH in the previously buffered batch. Precipitated proteins are separated on a solids-ejecting centrifuge and then washed by acidulated water and re-centrifuged. The resulting protein isolate is neutralized by sodium hydroxide and then spray-dried.

The mass balance from a continuous pilot-plant production of fababean and pea proteins is given in Table 8. Through this process, the results for fababean and pea protein extraction are very similar. In both cases, the distribution of the protein and the non-protein components follows approximately the same pattern. About 85% of the flour protein was extracted and 75% to 80% was precipitated, generally leading to a recovery yield of total initial protein of between 58% and 65%, and to protein isolates having a protein content of 90% to 95% (N x 6.25). As a rule, when the protein contents of the pea and fababean isolates were compared, the fababean isolate was 2% to 3% higher in protein than the pea isolate. From a technical point of view, the equipment used for the solid-liquid separation is suitable both for fababean and pea.

The biochemical composition of the isolates (Table 9) shows that most of the non-protein material was composed of lipids; the amount of which depended on the protein source. Fababean isolate had the lowest lipid content (5%), pea isolate reached 6% to 8% lipid and lupine isolate lipid levels were as high as 13.5% [36]. Only 0.3% ~~of the lipids of fababean~~ and 1.6% ~~of the pea lipids~~ were extractable by hexane, the remaining portions being bound to the protein. Most of lipids are free in the flour, but become attached to the protein during the processing. Indeed, only 2.5% of the lipids were found in the isolates prepared from defatted pea flour. The binding mechanism is not very well known but it could be due partly to an emulsification of the lipid by the protein. While extracting the pea protein,

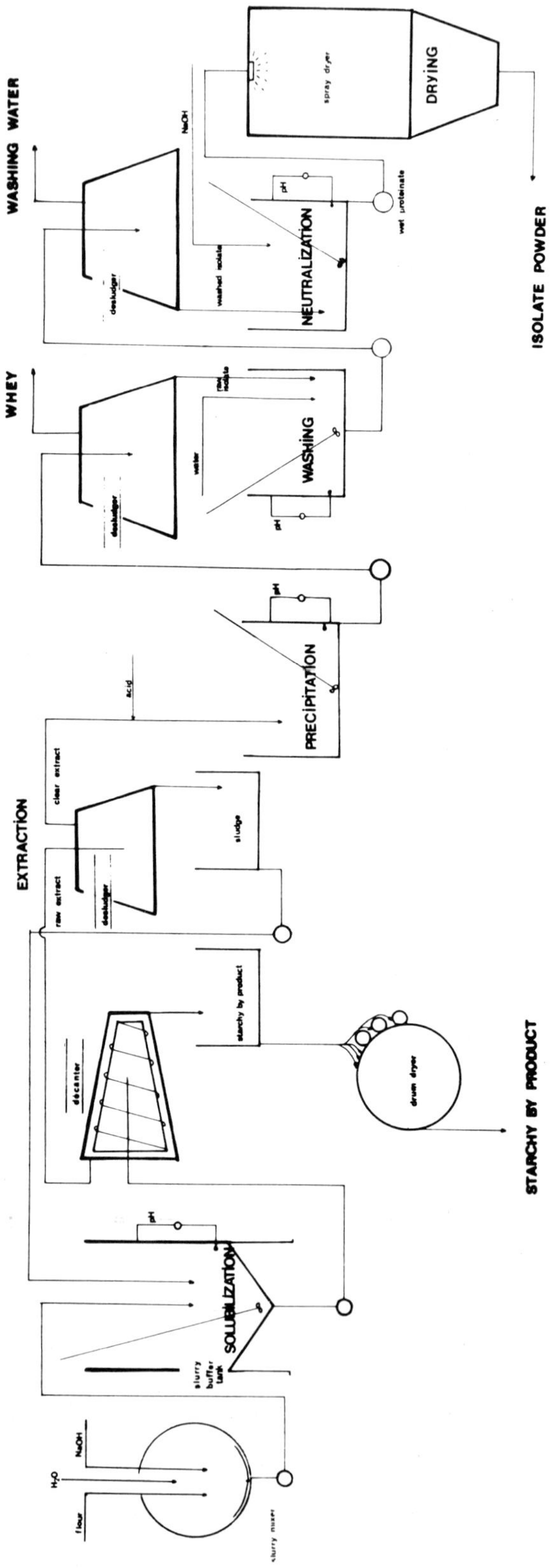

Figure 5. Flowsheet for fababean and pea isolate production.

Table 8. Mass and protein balance from continuous pilot plant production of pea and fababean protein

	Pea (Var. Frimas)			Fababean (Var. Ascott)		
	Yield of protein (N × 6.25) (%)	Yield of dry matter (%)	Ratio of protein to dry matter	Yield of protein (N × 6.25) (%)	Yield of dry matter (%)	Ratio of protein to dry matter
Raw material						
Flour	100	100	23	100	100	33
Protein solubilization						
Slurry	100	100	23	100	100	33
Separation						
Extract	87	35	57	84	43	70
Starchy by-product	13	65	5	16	57	10
Precipitation						
Raw protein isolate	62	17	84	68	26	86
Whey	25	18	31	16	17	31
Washing						
Washing water	2	1.5	30	1.5	2	25
Protein isolate	60	15.5	90	66.5	24	92

Table 9. Biochemical composition of pea, fababean and lupine flours and protein isolates[a]

	% d.m.	Protein (N × 6.25)	Lipid	Ash
Pisum Sativum (Smooth pea; var. Amino)				
– Flour	88.7	23.4	1.7	3,2
– Isoelectric protein isolate				
ext. pH 9, prec. pH 4.5	93.5	90.1	8.5	0.9
ext. pH 7, prec. pH 5.3	96.5	93.0	6.2	2.1
– Proteinate				
ext. pH 9, prec. pH 4.5	95.7	89.0	8.0	5.9
Vicia faba (Var. S 45)				
– Flour	87.7	32.5	2.0	3.87
– Isoelectric protein isolate				
ext. pH 9, prec. pH 4.5	95.3	91.2	5.0	2.6
ext. pH 7, prec. pH 5.3	92.7	95.6	3.2	2.2
ext. pH 7 (U.F.), prec. 5.3	92.9	96.2	4.2	1.7
Lupine[b]				
– Flour	90.3	34.7	10.4	–
– Isoelectric protein isolate	94.3	75.6	13.5	3.4
– Ultrafiltered protein isolate	95.7	72.3	13.8	3.7

[a]Results expressed as % of dry matter; lipids were extracted after acid hydrolysis (method of Drapron [14]); ext. = extracted; prec. = precipitated. U.F. = ultrafiltered extract.
[b]From Pompéi and Lucisano [36].

the lipid content of the isolate can be increased from 6% to 8.5% at pH 7 by a vigorous mixing of the slurry.

Because of the effects of lipids on the functional properties and flavours of isolates, some experiments have been carried out to decrease the lipid content. A pH step precipitation procedure has been used. The protein fraction recovered between pH 9 and pH 7 contains a very high lipid (20%) and a lower protein (about 50%) content for pea as well as for fababean, and represents 7% of the dry matter solubilized in the whole extract. Moreover, the protein content of the whole acid precipitated isolate decreases when the precipitation pH is chosen below pH 5.3. Therefore, a process using pH 7 and pH 5.3 for extraction and precipitation conditions has been developed. Under these conditions the protein recovery yield was reduced but remained between 50% and 55% while the lipid content was about 25% lower. This process appears to be potentially more useful for fababean than for peas, since the lipid contents were about 3% and 6%, respectively (Table 9).

The importance, on a weight basis of the starchy by-products, their high starch contents (78.5% and 78.7%, respectively for fababean and peas [9]), and the interesting properties, in general, of legume starches led us to purify them by sieving and centrifugation. Two procedures have been used [10]. For fababean, the first process (A) consists of dispersing the flour in water at pH 9 with a 1:10 (w/w) ratio of flour:water, and then screening the slurry through a bank of 200, 160, 125, 100, 80 and 60 vibrating screens. In the second process (B), the by-product resulting from the protein extraction is once again suspended in water (ratio 1:50) and then sieved on the same vibrating screens. For the pea, this latter technique has only been tested with two by-products:water ratios (process C, ratio 1:50; process D, ratio 1:10). In all cases, the starch is recovered from the underflow fraction by centrifugation with the cell-wall material-rich fractions remaining on the screen.

The similarity between the starch-fraction composition obtained from fababean flour by the A and B processes (Table 10) indicates that the removal of protein by centrifugation before screening does not improve their purity. However, this process leads to a loss of the protein material because of the water imbibed by the cell wall material. These fractions retain 10 to 20 times their weight of water. Thus, it seems necessary to carry out the sieving after the protein removal by centrifugation. The ratio of by-product:water was observed to be important in peas. When this ratio was 1:10 instead of 1:50, the residual starch retained in the cell wall material increased. Starch recoveries were 95.2%, 92.1%, 97.7% and 94.1%, respectively, for tests A, B, C and D. The chemical composition of these starches was characterized by small amounts of cell wall material (less than 4%) and minor lipid and protein contents. Microscopic observations showed that they were not degraded. The fiber fractions were mainly concentrated on the large size screens (200 μ to 125 μ). They are characterized by a very high water absorption capacity and still contain small amounts of protein (4%–8%) and lipids (0.5%–1.5%) [10].

Concerning the protein extraction conditions and recovery yields, these results are consistent with previous work [18, 30, 31, 34, 45]. The biochemical composition of the isolates did not differ from the results recently published by Vose [53], who also noticed a high lipid content in the isolates but found more lipid in fababean isolate (6.3%) than in pea isolate (4.9%).

The protein recovery yield in our process (about 60%) is lower than that found by Vose (77%) [53]. This phenomenon results from differences in the starch separation process. In Vose's procedure, the flour slurry is screened initially on a 75 μm screen to remove the fiber fraction and the starch and the protein extract in the underflow are separated through five hydrocyclone units. This operation, which involves the use of a countercurrent water wash system and leads to a very efficient extraction of protein from the starchy fraction, explains the 77% protein recovery. In fact, it is possible to reach this value in our process by recirculating the washing water of the starchy by-product separated from the purified starch by centrifugation.

A simpler wet process, derived from those described formerly, has also been developed for processing pea protein by the Prairie Regional Laboratory (Canada) [37]. In the process, called the slurry centrifugation process by the authors, the whole alkaline extract was spray-dried. But the flour was

Table 10. Composition of fababean and pea starch fractions by a sieving process[a]

	% Ins.[b]	Protein N × 5.8	Starch	Furfural generator	A.D.F	Ash	Ethanol soluble sugars	Lipids
Vicia Faba								
Flour		25.9	56.3	3.0	1.2	3.4	8.4	2.1
Proc. A								
Flour (1:10)	81.0	0.5	96.7	1.7	0.2	0.3	*	0.3
Proc. B								
by-product (1:50)	84.2	0.4	96.2	1.8	0.2	0.3	*	0.2
Pisum Sativum								
Flour		24.5	51.4	2.9	1.9	3.2	7.7	2.0
Proc. C								
by-product (1:50)	91.7	0.3	93.8	3.1	0.5	0.4	*	0.2
Proc. D								
by-product (1:10)	89.6	0.3	94.8	2.5	0.3	0.3	*	0.2

[a]From Colonna et al. [10].
[b]% Ins. = starch fraction as percentage of the insoluble by-product fractions.

suspended in only five parts of water. In this way, a pea protein concentrate was obtained which had a protein content of 60%.

The process developed at the Food Technology Laboratory (Lyngby, Denmark) by Olsen [30, 31, 34] on fababean used ultrafiltration instead of acidificiation to recover the extracted proteins; the overall recovery yield (62%) and the protein content of the spray-dried isolate (90%) are close to our own results. No details were given on the lipid content. Vose [53], using a similar process, found 6.1% lipid in pea isolate but only 1.6% lipid for fababean ultrafiltered isolate (no explanation for this difference was given). Such a difference in the lipid content of an isoelectric precipitated isolate compared to an ultrafiltered isolate has not been observed in our experiments on fababean (Table 9); lipid contents of 3.2% and 4.2% have been obtained on both products.

The main problem preventing the use of ultrafiltration in recent years has been the low flow rate and the plugging of the membrane when the protein concentration increases [30]. However, Olsen claimed that a satisfactory flow rate can be obtained by using ultrafiltration moduli under well defined conditions. Firstly, the protein extract has to be heated at 50°C, in order to decrease its viscosity. Secondly, the flow rate, which depends significantly on the membrane capacity, is also modified to a large extent by the dry matter concentration in the extract. According to this author, only the second generation type of membrane and modulus is satisfactory. These membranes are made of a modified polysulphone and the design of the module allows a high flow velocity and a low pressure drop over the channels. Even with this equipment, a protein cake can form on the membrane if the flow rate is too low (< 2.5 m/s). However, in these well defined conditions, with a batch process, a fababean protein extract can be concentrated from 5.7% to 24.9% dry matter at an average membrane capacity of $45.5\,l/h/m^2$. In contrast, with a first generation modulus the same results are only obtained with a membrane capacity of $14.7\,l/h/m^2$. An ultrafiltration two-step continuous plant was also developed by Olsen who observed that the membrane capacity plotted against the average pressure shows a very sharp maximum at $2.80\,kp/cm^2$ [31]. Moreover, the second generation membrane capacity is only $7.1\,l/h/m^2$ instead of $45.5\,l/h\,m^2$ with the batch process. The volume and mass balance per hour are given in Figure 6 and Table 11 and confirm the lower efficiency of the continuous process. Even in the optimal conditions used by Olsen [31], with the continuous process the flux is gradually reduced because of the inadequate removal of the protein gel layer which is formed either on or inside the membrane surface.

These technical difficulties make it very hazardous to use ultrafiltration on an industrial scale, although Olsen claimed that the operating cost is the same for the continuous ultrafiltration or the acid precipitation processes. In fact, even if the scale-up is not very easy to achieve, this process remains very interesting as a complementary tool of the isoelectric technique. This

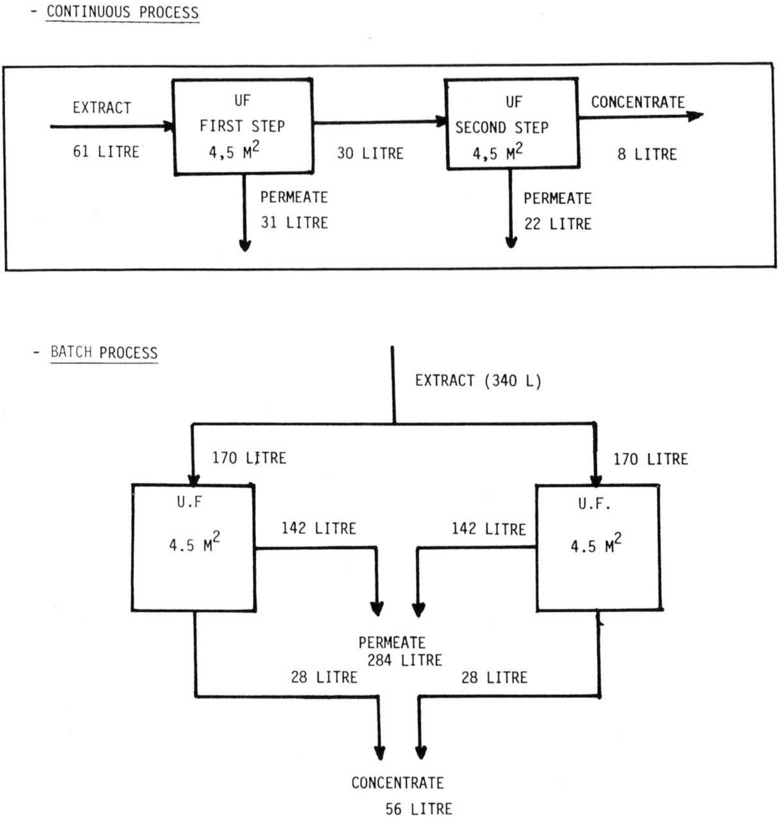

Figure 6. Comparison of the volume balance per hour for continuous and batch ultrafiltration processes of fababean extract, giving the same protein: dry matter ratio in the concentrate; from Olsen [31].

Table 11. Comparison of the protein and dry matter composition of concentrate and permeate from continuous and batch ultrafiltration of fababean extract[a]

	First step			Second step	
	Extract	Concentrate	Permeate	Concentrate	Permeate
Continuous					
% Dry matter (d.m)	6.5	10.8	2.1	26.8	2.7
% Protein (P)	4.7	9.2	0.7	24.3	0.9
% P:% d.m.	72	85	33	91	33
Batch					
% d.m.	5.7	24.9	0.6		
% P	4.1	22.3	2.4		
% P:% d.m.	72	90	25[b]		

[a]From Olsen [31].
[b]Calculated from the results of Olsen [31].

method could indeed by very valuable because of the high capacity of the new membranes to prepare proteins which are not denatured by acid precipitation.

Furthermore, an ultrafiltration and diafiltration batch process has been tested by Pompei and Lucisano [36] to concentrate a lupine protein extract. The ultrafiltered isolate has a protein content of 72.3% which is lower than the acid precipitated isolate and a lipid content of 13.8%. From these results it also seems that ultrafiltration does not decrease the lipid content of the isolate.

A newly issued Canadian patent [28] describes a method for the preparation of isolates by extracting the protein with a salt solution (NaCl) of at least 0.2 ionic strength. The proteins are precipitated as micelles when the ionic strength is decreased by dilution with water. Examples are given, especially for pea and fababean flour and concentrates. It is claimed that an isolate with a very high protein content (95%) can be obtained. No information is given either on the lipid content of these isolates or on the recovery yields.

Product characteristics

Antinutritional factors

These different technologies, air-classification, acid precipitation and ultrafiltration lead to products which contain various amounts of antinutritional factors. By air-classification, most of the antinutritional factors are recovered in the light fraction. Vose et al. [52], in experiments with peas, observed that a large amount of α-galactosides are concentrated in the protein fraction. On the other hand, no significant distribution difference has been seen with the total soluble sugars of the fababean, whereas more verbascose and stachyose has been found in the fine fraction (Table 12). Trypsin inhibitors and hemagglutinins, which follow the same distribution as the other proteins through air-classification, are generally more and more concentrated in the light fraction as its protein content increases (Table 13).

The amounts of glucosides, vicine and convicine of fababean are also higher in the protein fractions than in the starch ones. Pitz et al. [35] have shown that in the fine fraction the total glucosides content is about 2% although the heavy fraction contains only 0.6% after the first run and 0.3% after the second (Table 14). The isolates contain less antinutritional factors than the concentrates. Gueguen et al. [22] showed that the α-galactosides of fababean and pea are solubilized in the protein extract but remain in the whey during the isoelectric protein precipitation (Table 15). By washing the isolate with water, the residual part remaining in the wet protein curd is eliminated. These oligo-saccharides can also be removed from the protein extract by ultrafiltration because of their low molecular weight which allows

Table 12. Partition of oligosaccharides between the different fractions obtained by air-classification of dehulled, pin-milled legume flours[a]

	Sucrose	Raffinose	Stachyose	Verbascose	Total
Pea					
Flour	2.2	0.8	2.0	2.3	7.4
P_I	2.2	1.1	4.0	3.9	11.2
S_I	1.9	0.6	1.3	1.4	5.2
Fababean					
Flour	2.9	0.2	0.2	2.0	5.6
P_I	1.6	0.2	0.8	3.2	5.8
S_I	3.4	0.2	0.4	1.5	5.5

[a]From Vose et al. [52]; values given are % dry matter.

Table 13. Trypsin inhibitor and haemagglutinin activities of fababean flour and concentrates[a]

	Trypsin inhibitor (TUI/mg dry matter)	Hemagglutinating activity (units/mg dry matter)
Flour	6.9	100
Concentrates		
40% protein	12.3	100
60% protein	14.9	200

[a]From Valdebouze and Melcion [47].

Table 14. Vicine and convicine contents of fababean fractions obtained by pin-milling and air-classification[a]

Fraction	Protein content (N × 6.25)	% glycosides in fraction (dry basis)		
		Convicine	Vicine	Total
Whole seeds (meal)	28.1	0.26	0.67	0.93
Flour	31.3	0.29	0.68	0.97
Starch I	21.3	0.19	0.47	0.66
Starch II	12.4	0.08	0.23	0.31
Protein I	71.5	0.68	1.50	2.18
Protein II	66.5	0.63	1.45	2.08

[a]From Pitz et al. [35]; see Vose et al. [52] for procedures used to obtain starch and protein fractions I and II.

them to flow through the membranes along with the permeate. Ultrafiltered or acid precipitated isolates contain only slight traces of these flatulence factors. Olsen and Andersen [32] observed a similar phenomenon with vicine and convicine in fababean. The glucosides content of isolates, contrary to the air-classified protein fraction, is always smaller than the initial content of the flour. Acid precipitated isolate contains about 0.10% and 0.05% vicine and convicine, respectively, while in the ultrafiltered isolate, the corresponding amounts of these glucosides is about 0.33% and 0.23%. Only

Table 15. Partition of the oligosaccharides between the intermediary and end products resulting from pea and fababean protein isolate processing[a]

		Sucrose	Raffinose	Stachyose	Verbascose	Ajugose	Ethanol soluble sugars
Vicia Faba							
	Flour	2.00	0.20	0.95	2.90	+	5.90
Laboratory	Extract[b]	1.94	0.20	1.03	3.20	+	6.00
Laboratory	Whey[b]	1.94	0.19	1.05	3.30	+	6.10
Laboratory	Non-washed isolate	0.80	0.12	0.15	0.78	Traces	1.90
Pilot Plant	Washed isolate	0.02	0.01	$< 5\ 10^{-4}$	Traces	Traces	–
Pilot Plant	Washed-isolate Precipitated from U.F.	$< 5\ 10^{-4}$	$< 5\ 10^{-4}$	$< 5\ 10^{-4}$	$< 5\ 10^{-4}$	Traces	–
Pisum Sativum							
	Flour	2.35	0.60	1.70	2.60	+	6.30
Laboratory	Extract[b]	2.28	0.62	1.75	3.08	+	7.10
Laboratory	Whey[b]	2.30	0.62	1.75	3.05	+	7.20
Laboratory	Non-washed isolate	0.85	0.25	0.85	1.30	+	2.70
Pilot Plant	Washed isolate	0.04	0.02	0.15	0.20	Traces	–

[a]From Gueguen et al. [22].
[b]Expressed in mg/ml for extract or whey prepared from 100 g of dry flour.

a diafiltration process can lead to a very low content (0.057% and 0.019%, respectively, for vicine and convicine). In fact, isolated proteins from faba-bean have a glucosides content which depends on the degree of separation and washing of the protein extract or isolate.

In the case of trypsin inhibitor and hemagglutinating activities, the elimination was not as successful. Gueguen et al. [22] (Table 16) have shown that about a third of the initial trypsin inhibitor and hemagglutinating activities of the flour are still present in the isolate, the remaining two-thirds being eliminated in the whey. Varying the parameters of the process can also decrease the amount of these components in the isolate [22] (Table 17). The extraction pH has a slight effect on the trypsin inhibitor content of the isolate. Trypsin inhibitors, with their isoelectric pH in the range of pH 9, are probably less soluble when protein extraction is carried out at this pH instead of pH 7. Moreover, since the overall protein recovery yield is higher at pH 9, the dilution of the trypsin inhibitor is greater in the end product. These phenomena probably explain the slightly lower trysin inhibitor content of pH 9 extracted isolates compared to the pH 7 extracted ones. For hemagglutinating activity, the influence of the extraction pH is less significant. The drying conditions, including spray-drying, do not modify the amount of these antinutritional factors in the isolate. Only the ultrafiltration of the extract before precipitating the protein leads to a lower content, mainly of the trypsin inhibitors which decreased from 7.8 TUI/mg to 2 TUI/mg. This result is not comparable to Olsen's data [32] which show that the ultrafiltered fababean protein isolate is richer in trypsin inhibitors than the acid-precipitated one. This difference probably derives from the fact that in Olsen's process the whole ultrafiltered concentrate is spray-dried, whereas in our case the proteins are acid precipitated from the ultrafiltered concentrate. Thus, as shown above, most of the trypsin inhibitors are left in the whey.

However, in all cases, the trypsin inhibitor contents of fababean and pea isolate or concentrate reach similar mean values as in the commercial soybean products. In promine D, 40.6 TUI/mg of protein was found [33]. Kakade et al. [25] indicated 8.5 and 19.8 TUI/mg of sample for commercial soybean isolate, and 12.0 TUI/mg for soy protein concentrate.

Few data are available on the phytic acid content of isolates and concentrates. According to Vose et al. [52], dehulled fababean may contain as much as 1.8% phytate and the protein concentrate as much as 4.2%. In our own pea products, the amount of phytic acid has been found to be 0.96% in the dehulled seed and 0.9% for the isolate extracted at pH 7 and precipitated at pH 5.3, but this increases to 2.1% when the alkaline extract at pH 9 is acidified at pH 4.5 (assuming the empirical formula $P_6O_{24}H_{18}$). Very similar results were described by De Rham and Jost [13] for soybean isolates with 1.84% and 0.60% in isolates precipitated from alkaline extract (pH 8.2) at pH 4.5 and 5.5, respectively.

Table 16. Partition of trypsin inhibitor and haemagglutinating activities between the intermediary and end products resulting from pea and fababean protein isolate processing[a]

	Vicia Faba			*Pisum Sativum*		
	Trypsin inhibitor[b]		Hemagglutinating[c] activity (units/mg d.m.)	Trypsin inhibitor[b]		Hemagglutinating[c] activity (units/mg d.m.)
	T.U.I./mg d.m.	Flour T.U.I. (%)		T.U.I./mg d.m.	Flour T.U.I. (%)	
Flour	5.4	100	50	8.4	100	205
Starchy by product	0.13	9.6	6	1.2	8.5	25
Whey	20.8	56.4	50	43.2	73.2	410
Non-washed isolate	5.8	28.7	25	3.0	5.4	25
Total		94.7			87.1	

[a]From Gueguen et al. [22].
[b]T.U.I. = trypsin units inhibited (Kakade et al. method [25], slightly modified).
[c]Hemagglutinating activity (Valdebouze method [48]).

Table 17. Influence of processing conditions on trypsin inhibitor and haemagglutinating activities of fababean and pea flours and protein isolates[a]

	Extraction pH	Processing conditions	Trypsin inhibitor[b] (T.U.I./mg dry matter)	Hemagglutinating activity[c] (units/mg dry matter)
Vicia Faba				
Flour			7.8	50
Isolate 1	pH 9	Washed, freeze-dried	6.0	13
		Washed, spray-dried	4.5	25
Isolate 2	pH 7	Washed, freeze-dried	7.8	6
Isolate 3	pH 7	I.E. isolate from an U.F. extract: Washed, freeze dried	2.0	6
Pisum Sativum				
Flour			9.7	200
Isolate	pH 9	Washed, freeze dried	5.7	100
		Washed, spray dried	5.6	100

[a]From Gueguen et al. [22].
[b]T.U.I. = trypsin units inhibited (Kakade et al. method [25], slightly modified).
[c]Hemagglutinating activity (Valdebouze method [48]).

These results, concerning the presence of antinutritional factors in extracted protein products from fababeans and peas, allow for the generalization to be made that concentrates generally contain higher concentrations of antinutritional factors than either isolates or flours. The difference between ultrafiltered and acid precipitated isolates is slight. For isolates as well as for concentrates, the amount of these components in fababean and pea protein-rich products were in the same range and often less than in the homologous soybean products. However, more nutritional studies are needed to determine whether or not it is still necessary to decrease the level of these antinutritional factors by improving the different processes.

Functional properties

Spray-dried pea and fababean isolates obtained from dehulled seed flour are generally cream coloured and quite bland in flavour. Pea isolates are judged to be as bland as the corresponding soy isolate by a panel [30]. Lupine isolates are more yellow. Variations in the functionality of the end products can be due to the raw material used and to the process. Sodium proteinate isolates exhibited as a rule more functionality than the isoelectric ones.

Pea and fababean isolates have nitrogen solubility curves which are very similar to the native seed protein. The nitrogen solubility is lower in the isoelectric range (pH 4.5) but reaches 66% at pH 7 and pH 3 for an isoelectric precipitated pea isolate (by A A C C method). Results with fababean are very similar (Figure 7) [38, 53]. According to Quaglia et al. [38], lupine isolates seem more soluble with their solubility reaching 80% at both pH 7 and pH 3. Pea protein isolates are generally more soluble than the corresponding soybean isolates, particularly if the isoelectric ones are compared [45]. At pH 7, the nitrogen solubility of the pea isoelectric isolate (evaluated by the A O C S method), reached 72% whereas it is only 30% for soybean. In contrast, Sumner et al. [45] did not find this difference between either soya or pea sodium proteinate isolates.

The water retention capacity of pea and fababean is estimated to be 2.7–2.8 times their weight of water at pH 7. This level is lower than those generally observed with soybean isolates which absorbed 4–5 times their weight of water [53].

The thickening properties of our sodium pea proteinate compared to a homologous soybean product have also been studied [23]. The rheological properties of aqueous dispersions of the isolate have been examined as a function of the protein concentration. Flow curves between 0 and 698 s^{-1} can be described with three parameters σ_0, K and n using the power low $\log(\sigma - \sigma_0) = \log K + n \log \dot{\gamma}$ (σ: shear stress, σ_0: yield-stress, K: consistency index, n: behaviour index). They show that the behaviour of the dispersion is non-Newtonian and time dependent (partial thixotropy). Comparing pea and soybean isolates, the protein concentration has to be higher in pea protein dispersions than for soybean to obtain dispersions with the

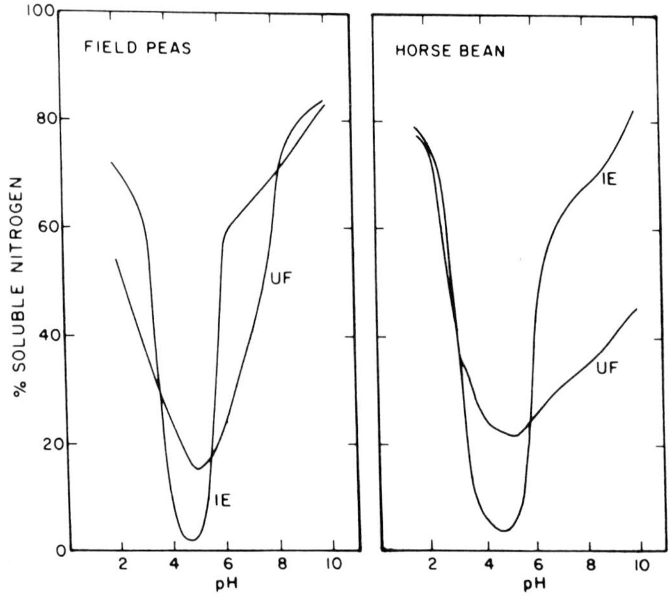

Figure 7. Effect of pH on the nitrogen solubility of spray-dried isolates produced by isoelectric precipitation or ultrafiltration from peas and fababean (horse bean); from Vose [53].

same viscosity. The consistency index K reaches 210 dyn/cm^2 with a 15% soybean proteinate dispersion whereas a concentration of 26% is needed for peas (Figure 8).

Concerning emulsifying properties the results obtained with pea and fababean isoelectric and ultrafiltered isolates are comparable to those obtained with the soybean isolates (e.g., Supro 620 [53]; Figure 9). Even in a 4% NaCl solution, the emulsifying activity of both pea and fababean isolates is higher. The foaming stability and expansion of all of these isolates have also been compared (Table 18; Figure 10). The ultrafiltered fababean isolate shows the best foaming properties. However, compared to commercial soybean isolates (Promine F and Supro 620), all of the fababean and pea isolates are found to be generally equivalent or superior.

Spinning experiments have also been carried out successfully on our isoelectric or proteinate isolates for fababean as well as for pea [11, 16]. The fibers produced are cream coloured and look very similar to spun soybean fibers.

Functional properties can vary as a function of the operating conditions used for processing. Thus, isolates prepared by ultrafiltration have higher solubilities compared to acid precipitated products, especially in the isoelectric pH range [3, 53]. But from a comparison of the results of Olsen [31] and Vose [53] on fababean, it seems that the operating conditions can be

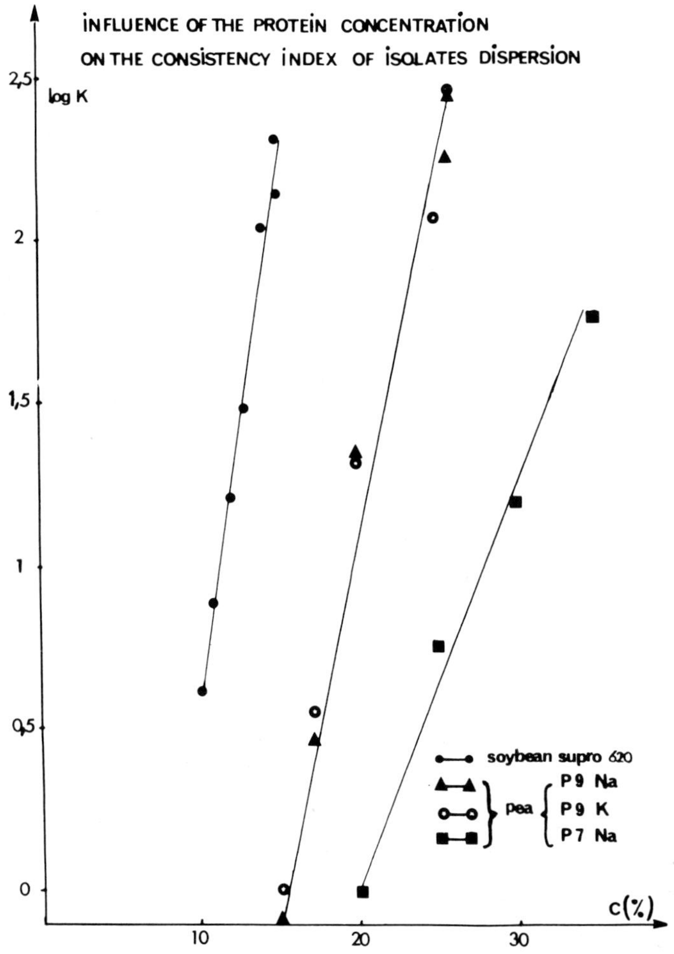

Figure 8. Influence of the protein concentration on the consistency index of isolates dispersion; from Gueguen and Lefèbvre [23].

used to modify the nitrogen solubility profile. The ultrafiltrated isolate of Olsen is more soluble in the alkaline pH range, whereas with Vose's data, the reverse is true. The extraction pH (7 or 9) and the precipitation pH can also influence the flow properties of the protein dispersion and their solubility characteristics [23]. A pH 7 protein dispersion obtained with a pH 7 extracted pea isolate at a concentration of 25% (w/v) gave a consistency index of 5.6 dyn/cm^2, instead of 221 dyn/cm^2 which was observed when a pH 9 extracted proteinate was used (Figure 8). Drying also seemed to have very important effects. Comparing different dry processes on isoelectric and sodium proteinate pea isolate, Sumner et al. [45] observed that drum-drying

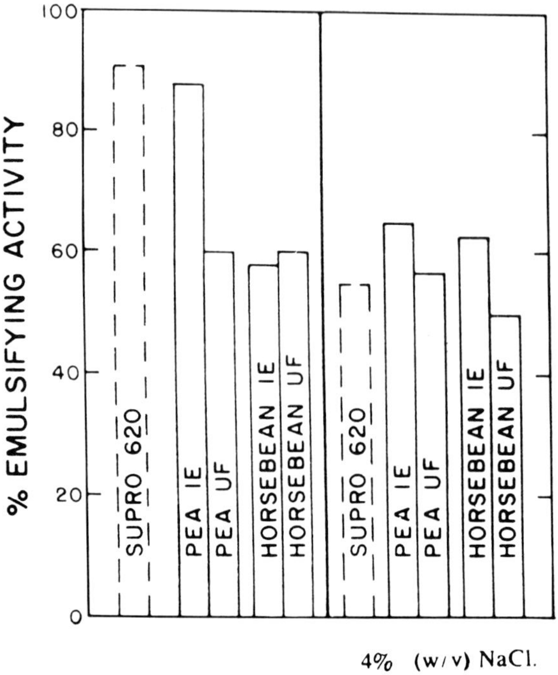

Figure 9. Emulsifying properties of legume protein isolates including soybean (Supro 620), pea, and fababean (horsebean) isolates; from Vose [53].

gives darker products than spray-drying or freeze-drying which are characterized by a low nitrogen solubility, and higher water adsorption and emulsification capacities. Freeze- and drum-dried isolates exhibit the highest fat adsorption whereas spray-drying produced the best foaming, colour and flavour properties.

Air-classified concentrates generally exhibit less functionality than isolates, especially for foaming and emulsifying properties (Table 18), but the presence of the residual starch in the protein fraction can be useful in relation to viscosity and binding properties. Comparing eight legume protein concentrates, Sosulski and Youngs [44] showed that northern bean had better functional properties.

Conclusions

From a technical point of view, these studies have demonstrated that European legume seeds such as peas and fababeans can be used as protein sources for isolate and concentrate production. Less work has been devoted to lupine protein extraction, probably because of the high lipid content of the seed which leads to isolates with a low protein and high lipid content.

Table 18. Foam characteristics and emulsifying activities of fababean air classified concentrates[a]

	Protein content (% N X 6.25)	N.S.I.	Foam[b] Expansion (ml)	Foam[b] Stability (ml)	emulsifying[b] activity (%)
Fababean					
Concentrate	50	70	15	8	66
Concentrate	65	70	22	20	69
Na proteinate		N.D.	40	31	
Soybean					
Commercial isolate	91	56	40	32	83

[a]From Vallery-Masson et al. [50].
[b]Method of Yasumatsu et al. [54].

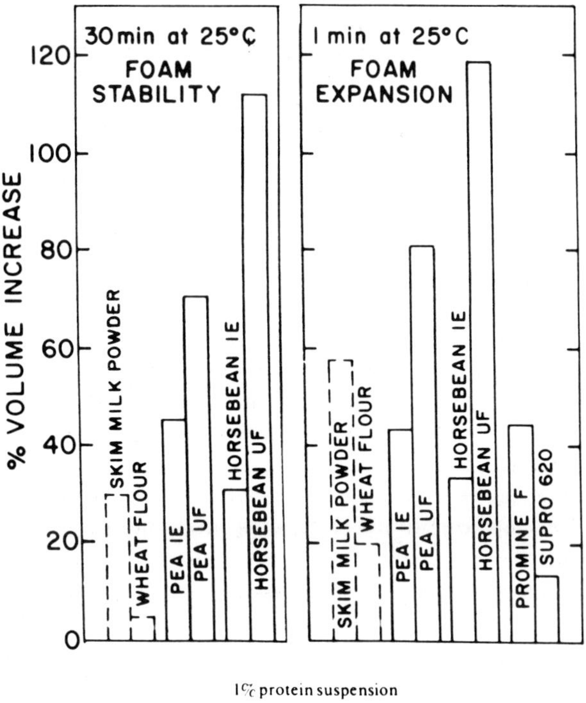

Figure 10. The foaming properties of legume protein isolates including soybean (Promine F, Supro 620), pea, and fababean (horsebean) isolates and of other protein sources; from Vose [53].

The air-classification process is very well adapted to both pea and fababean flours, due mainly to the quite large diameter and fairly uniform distribution of their starch granules. From the Canadian work, it appears that this technique can be applied successfully to several other starch-rich legume crops, such as those grown in developing countries, but the technique does not give satisfactory results for lipid-rich seeds like lupine, wrinkled pea, chick-pea or soybean. Wet processes, including acid precipitation as well as ultrafiltration, have also been shown to be technically easy to develop. The products of these processes do not contain more antinutritional factors than homologous soybean products and they exhibit comparable and even, to some extent, complementary functionality. In this respect, air-classified products are different from soybean concentrates because of their residual starch content which can be very useful for particular functional applications. Pea and fababean isolate functionality behaviour also seems to be partially different from that of soybean isolates. The pea and fababean isolates have better foaming properties and are generally more soluble, but they have to be more concentrated to produce highly viscous dispersions. Further work has to be done in this area to determine the most useful specific applications of these new isolates compared to those of soybean. Studies will also be necessary to evaluate the marketability of these new products. From an economical point of view, it appears from the French experiment by Grandes Minoteries à Fèves that pin-milling and air-classification can be used successfully on fababean. For pea protein processing, it would be probably more difficult because of the lower protein content of the flour. According to Bérot and Briffaud [4], the economic balance of the extraction processes is closely related to the protein content of the raw material. Consequently, economic evaluations of wet processes have been carried out on air-classified pea protein fractions which have increasing protein contents. The results obtained have shown that in every case the pre-concentration cost was too high to improve the economic balance sheet of the process.

In conclusion, the feasibility of producing pea and fababean isolates is related (a) to the particular functional properties of the isolates compared to soybean isolates and (b) to the marketability of the by-products. For these reasons, in addition to the studies carried out on the isolate processing and on the physiochemical and functional properties of fababean and pea proteins, research programmes have been developed by Canadian teams and are being carried out today in our research center on the composition of legume starches and on the functional and nutritional properties of both protein isolates and polysaccharides-rich fractions separated from the starchy by-products of the protein extraction.

Acknowledgements. Thanks are due to Mrs. L. Joly, Dr. C. Mercier and M.J. Lefebvre for their help in writing the text in English.

References

1. Aman P (1979) Carbohydrates in raw and germinated seeds from mung bean and chickpea. J Sci Food Agric 30:869–875

2. Anson ML, Pader M (1957) Extraction of soy protein U.S. Pat. 2 785 155
3. Ballester D, Yanez E, Garcia R, Erazo S, Lopez F, Haardt E, Cornejo S, Lopez A, Pokniak J, Chichester CO (1980) Chemical composition, nutritive value and toxicological evaluation of two species of sweet lupine *Lupinus albus* and *Lupinus luteus*, J Agric Food Chem 28:402–405
4. Bérot S, Briffaud J (1980) Personal communication
5. Boulter DA (1977) Quality problems in 'Protein plants' with special attention paid to the proteins of legumes. Protein quality from leguminous crops CEE EUR 5686 EN:42
6. Cegla GF, Bell KR (1976) High pressure liquid chromatography for the analysis of soluble carbohydrates in defatted oilseed flours. J Am Oil Chemist's soc 54:150–152
7. Cerning J, Saposnik A, Guilbot A (1975) Carbohydrate composition of horse beans *Vicia faba* of different origins. Cereal Chem 52:125–138
8. Cerning-Beroard J, Filiatre A (1976) A comparison of the carbohydrate composition of legume seeds: horse beans, peas, and lupines. Cereal Chem 53:968–978
9. Colonna P, Gallant D, Mercier C (1980) *Pisum sativum* and *Vicia faba* carbohydrates: studies of fractions obtained after dry and wet protein extraction processes. J Food Sci 45:1629–1636
10. Colonna P, Gueguen J, Mercier C (1981) *Pisum sativum* and *Vicia faba* carbohydrates: Part III, Pilot scale preparation of starch and cell wall material. Sci Alim 1:415–426
11. Culioli J, Bouchet B, Gallant D (1980) Influence of denaturation on the spinning of pea proteins – European Symposium 'Progress in Food Engineering,' Milan 179–180
12. Derbyshire E, Wright DJ, Boulter D (1976) Legumin and vicilin, storage proteins of legume seeds. Phytochem 15:3–24
13. De Rham O, Jost T (1979) Phytate protein interactions in soybean extracts and low phytate soy protein products. J Food Sci 44:596–600
14. Drapron R (1975) Methode rapide de determination quantitative des lipides totaux dans les produits céréaliers Ann Technol Agric 24:117
15. Duranti M, Riccardi A, Restani P, Cerletti P (1981) Isolation and purification of proteins from lupine seeds and corn germ. European Symposium 'Progress in Food Engineering,' Milan
16. Fabre A, Ruand M, Magnat R (Société Rhone Poulenc) personal communications
17. Fan TY, Sosulski FW (1974) Dispersibility and isolation of proteins from legume flours. Can Inst Food Sci Technol J 7:256–259
18. Flink J, Christiansen I (1973) The production of a protein isolate from *Vicia faba.* Lebensm Wiss u Technol J 6:102–106
19. Galoppini C, Fiorentini R (1979) Concentrazione e isolamento dei costituenti proteici dei legumi. Convegno 'Prospettive delle proteaginose in Italia: aspetti nutrizionali, agronomici, techolocici ed economici della utilizzazione della leguminose da granella Perugia,' 30 maggio-1 giugno
20. Gatehouse JA, Croy RRD, McIntosh R, Paul C, Boulter D (1980) Quantitative and qualitative variation in the storage proteins of material from the EEC point field bean test. In: Bond DA (ed) *Vicia faba;* Feeding value, processing and viruses 173–190
21. Gueguen J (1980) Solubilité des protéines de la fèverole (*Vicia faba L*) et du pois (*Pisum sativum L*) dans les solutions salines Lebensm Wiss u Technol 13:156–163
22. Gueguen J, Quemener B, Valdebouze P (1980) Elimination des facteurs antinutritionnels de la fèverole *Vicia faba L* et du pois *Pisum sativum L* au cours de la préparation de isolats protéiques. Lebensm Wiss u Technol 14:72–77
23. Gueguen J, Lefèbvre J (1981) Functional properties of protein isolates from peas. European Congress 'Plant Protein for Human Food,' Nantes-France
24. Horan FE (1974) Soy protein products and their production. J Am Oil Chemist's soc 31:67A–73A
25. Kakade ML, Rackis JJ, McGhee JE, Puski G (1974) Determination of trypsin inhibition activity of soy products: a collaborative analysis of an improved procedure. Cereal Chem 51:376–382

26. Longe GO (1980-81) Carbohydrate composition of different varieties of cowpea *Vigna unguiculata.* Food Chem 6:153–161
27. Martensson P (1980) Variation in legumin: vicilin ratio between and within cultivars of *Vicia faba* var minor. In: Bond DA (ed), *Vicia faba:* Feeding value, processing and viruses 159–172
28. Murray ED, Myers CD, Barker LD (1978) Protein product and process for preparing same Patent No 102 8552
29. Millerd A, Thomson JA, Schroeder HE (1978) Cotyledonary storage proteins in *Pisum sativum* III, Patterns of accumulation during development. Aust J Plant Physiol 5:281–294
30. Olsen HS (1975) Isolation of bean protein by ultrafiltration In: Proceedings of the symposium on separation processes by membranes, Ion. exchange and freeze concentration in Food Industry' IUFOST and FEEC Paris A.6-1-A6-21
31. Olsen HS (1978) Continuous pilot plant production of bean protein by extraction, centrifugation, ultrafiltration and spray drying. Lebensm Wiss u Technol 11:57–64
32. Olsen HS, Andersen JH (1978) The estimation of vicine and convicine in fababeans *Vicia faba L* and isolate fababean proteins. J Sci Food Agric 29:323–331
33. Olsen HS (1980) A survey of recent developments in industrial processing of *Vicia faba* protein. In: Bond DA (ed), *Vicia faba:* feeding value, processing and vivuses, 233–235
34. Olsen HS (1981) Developments in industrial processing of *Vicia faba* protein in Europe. Fabis 3:13–15
35. Pitz WJ, Sosulski FW, Hogge LR (1980) Occurence of vicine and convicine in seeds of some Vicia species and other pulses. Can Inst Food Sci Technol J 13:35–39
36. Pompéi C, Lucisano M (1976) Le lupin *Lupinus albus L.* Comme source de protéines pour l'alimentation humaine II Production d'isolats par coagulation acide et par ultrafiltration–diafiltration 9:338 344
37. Prairie Regional Laboratory (N.R.C) and College of Home Economics (Univ. Saskatoon) (Canada) (1974) Pea flour and pea protein concentrate. PFPS Bulletin 1:3–4
38. Quaglia GB, Alessandroni A, Ricardi A (1981) Progetto finalizzato 'Ricerca di nuove fonti proteiche e di nuove formulaziono alimentari,' Proprieta funzionali della proteine 20–33
39. Quemener B, Mercier C (1980) Dosage rapide des glucides ethanolosolubles des graines de légumineuses par chromatographie liquide sous haute pression. Lebensm Wiss u Technol 13:7–12
40. Quemener B, Gueguen J, Mercier C, (1981) Determination of vicine and convicine in broadbeans by high pressure liquid chromatography
41. Reichert RD, Youngs CG (1978) Nature of the residual protein associated with starch fractions from air classified field peas. Cereal Chem 55:469–480
42. Rouanet JM, Besançon P (1979) Effets d'un extrait de phytohemagglutinines sur la croissance, la digestibilité de l'azote et l'activité de l'invertase et de la (Na^+–K^+)-APPase de la muqueuse intestinale chez le rat. Ann Nutr Alim 33:405–416
43. Smith AK, Circle SJ (1978) Chemical composition of the seed. In: Smith AK, Circle SJ (eds), Soybean: chemistry and technology. AVI PUB: 73
44. Sosulski F, Youngs CG (1979) Yield and functional properties of air-classified protein an starch fractions from eight legume flours. J Am Oil Chemist Soc 56: 292–295
45. Summer AK, Nielsen MA, Youngs CG (1981) Production and evaluation of pea protein isolate. J Food Sci 46:364–372
46. Thank VH, Shibasaki K (1976) Major protein of soybean seeds. A straightforward fractionation and their characterization. J Agric Food Chem 24:1117–1121
47. Valdebouze P, Melcion JP (1977) Action concertée feverole, Contrat DGRST no 74 7 10 51
48. Valdebouze P, Bergeron E, Gaborit T, Delort-Laval J (1980) Content and distribution of trypsin inhibitors and hemagglutinins in some legume seeds. Can J Plant Sci 60:695–701
49. Vallery-Masson D, Lallemant J, Achache S. Personal communication

50. Vallery-Masson D (1980) Feves et feveroles, adjuvant traditionnel de panification, source de protéines pour l'alimentation humaine et animale. Ind Agr Alim 6
51. Varner JE, Schidlovsky G (1963) Intracellular distribution of proteins in pea cotyledons. Plant Physiol 38:139–144
52. Vose JR, Basterrechea MJ, Gorin PAJ, Finlayson AJ, Youngs CG (1976) Air classification of field peas and horsebean flours: chemical studies of starch and protein fractions. Cereal Chem 53:928–936
53. Vose JR (1980) Production and functionality of starches and protein isolates from legume seeds (Field peas and horsebeans). Cereal Chem 57:406–410
54. Yasumatsu K, Sawada K., Moritaka S., Misaki M, Toda J, Inada T, Ishii K (1972) Whipping and emulsifying properties of soybean products. Agr Biol Chem 36: 719–727

Food grade oilseed protein processing: sunflower and rapeseed

L. TRANCHINO, R. COSTANTINO, G. SODINI

ASSORENI, ENI Group Research Association, 00015 Monterotondo, Rome, Italy

Key words: protein, sunflower, rapeseed, extraction processes, oil extraction, human food

Abstract. Industrial sunflower and rapeseed meals cannot be directly used as a food because of their high fiber content and because of the presence of some undesirable constituents (such as hulls, polyphenolic pigments, etc.) or precursors of toxic compounds (glucosinolates, etc.). Edible protein products (flours, concentrates, and isolates) from these two sources can be obtained by carrying out, to various degrees, and with different procedures, extraction operations of non-proteic and potentially toxic or antinutritional components. All the possible combinations of the single extraction operations (removal of fiber, lipids, polysaccharides, etc.) were studied by various authors in order to develop an optimum process both from the economic and the product quality points of view.

In this report the problems related to the individual extraction operations, rather than to individual processes, are reviewed for reasons of recapitulation and to provide a common basis for comparison. Although it is impossible to reach a definitive conclusion it appears that some of the processes reviewed are able to produce very attractive raw materials for food manufacturing industries. Nevertheless, no food grade sunflower or rapeseed proteins have appeared on the market to date. However, sunflower and rapeseed protein sources will have to be taken into account in the near future as an added promising means for attacking food shortage problems.

Introduction

Sunflower and rapeseed rank, with soybean, cottonseed and groundnut, as the most important annual crops in the world which are grown for oil production (Table 1) [45].

The reasons for the development of these crops are mainly agronomic, but the yield and the quality of the oil and the meal contribute to it. Chemical and aminoacid compositions of these seeds are reported in Tables 2 and 3 [46, 48]. At present, sunflower and rapeseed meals are regarded as potentially important foodstuff ingredients. Recently both of these products have been investigated with the objective of exploring their use for providing nutritional protein for human consumption.

Industrial meal cannot be used directly as a food because of its high fiber content and because of the presence of some undesirable constituents (such as hulls, polyphenolic pigments, etc.), or precursors of toxic compounds

Table 1. Major oilseed crops: world production (1000 Metric Tons)[a]

Period	Soybean	Cottonseed	Groundnut	Sunflower	Rapeseed	Total oilseeds
1971–72	45,890	23,515	11,780	9,980	7,270	109,599
1972–73	49,879	24,545	10,210	9,640	6,800	111,271
1973–74	61,330	24,750	10,850	12,150	6,810	125,952
1974–75	54,570	23,360	11,485	10,735	7,440	120,356
1975–76	66,040	21,285	12,485	9,930	7,925	129,285
1976–77	59,910	22,330	11,115	9,970	6,920	120,820
1977–78	72,770	24,800	11,135	12,935	7,940	141,500
1978–79	78,110	23,560	11,995	13,075	10,720	148,575
1979–80	93,750	25,465	11,455	15,660	10,100	168,080
1980–81	81,520	25,790	11,060	13,210	11,240	154,300
1981–82	86,600	26,630	12,120	14,450	11,450	163,250

[a]Data from: ISTA, Hamburg, West Germany [45].

Table 2. Composition of sunflower seed and rapeseed

	Sunflower			Rapeseed		
	Seed	Kernel	Hull	Seed	Kernel	Hull
Weight proportion (%)	–	70–79	30–21	–	81–84	19–16
H_2O	8–10	5–6	11–12	10–13	6–8	11–14
Oil[a]	38–19	49–55	2–3	40–42	45–50	16–16.5
Protein (N × 6.25)[a]	20–27	27–30	4–5	23–28	27–30	15–17
Ash[a]	3–4	3–4	3–4	3–5	3–5	3–5
Crude fiber[a]	15–20	2–3	60–62	6–8	1.5–2.5	25–28

[a]Moisture-free basis.

Table 3. Amino acid composition of sunflower and rapeseed meals [46, 48]

Amino acid	Sunflower (grams a.a. per 16 g N)	Rapeseed (grams a.a. per 16 g N)
Arginine	8.5	5.6
Histidine	2.4	2.6
Lysine	3.6	5.8
Tyrosine	2.9	2.6
Phenylalanine	4.6	3.5
Cystine	1.8	2.4
Methionine	2.6	1.8
Serine	4.2	3.7
Threonine	3.8	3.8
Leucine	6.2	6.3
Isoleucine	3.9	3.7
Valine	4.8	4.8
Glutamic acid	21.4	16.6
Aspartic acid	9.6	6.2
Glycine	5.4	4.3
Alanine	4.1	3.9
Proline	5.1	6.4

(glucosinolates, etc.). In order to obtain edible protein products from these sources, it is necessary to remove part or all the other non-proteic components and, especially, to remove or inactivate the undesirable factors.

In this area, several research initiatives have been undertaken in various parts of the world by government supported laboratories (such as universities), research foundations and industrial laboratories.

In the following discussion, the problems associated with the purification of protein from sunflower seed and rapeseed, and their potential usefulness, are examined separately.

Sunflower seed

Due to the absence of toxic or antinutritional factors, sunflower protein products are promising food ingredients. The only problems may be:

- polyphenolic pigments such as chlorogenic acid, responsible for discoloration of sunflower meal and reduced bioavailability of some essential amino acids under certain conditions [10, 12, 62],
- hulls, which contain an exceedingly high lignin content for food grade products, and pigments which give undesirable coloration [14, 80]

The main non-proteic components of sunflower seeds are:

- lipids, generally almost completely extracted due to the high value of the oil that is obtained. Nevertheless, technologies for producing partially defatted products have been proposed recently [36],
- polysaccharides, non-proteic nitrogen and mineral salts, produced as residue during protein isolation.

The protein products from sunflower (flour, concentrates and isolates) are obtained by carrying out extraction operations of non-proteic components to various degrees and with different procedures. Figure 1 gives a summary of the various processes proposed for obtaining edible sunflower protein products. All of the possible combinations of the four fundamental operations (removal of fibers, lipids, polysaccharides, polyphenols) were studied in order to discover the best processes, both on a basis of economics and product quality.

Although it is impossible to reach a definite conclusion on this point, it is nevertheless possible to underline the principal advantages and disadvantages relative to the different processes. For reasons of recapitulation and to have a common basis of comparison, we carried out this study by analysing separately the fundamental operations rather than the processes:

Elimination of fibrous components

Hulls which are the major source of fiber components (Table 2) may be removed by using as a raw material:

(a) *The defatted flour,* with physical processes (sieving [57, 61], hydrocyclones [69], etc.) or chemical processes (selective extraction of protein with production of an isolate; discussed later).

At present, from the meal obtained by conventional industries, it is possible to obtain products with a higher protein content and a lower fiber level by means of mechanical operations, such as milling and sieving. The resulting products possess a nutritional value and an organoleptic character suitable for applications as protein integrators for animal feedstuffs.

However, only by excluding drastic thermal and mechanical treatments (pressing, toasting, etc.) in the defatting process is it possible to obtain food grade products. The protein, yield and content of the products obtained with various separation procedures following this line, are reported in Table 4. The highest quality product (flour with a high level of protein (60%), chemical score of 80; PER of 2.19; creamy white color; no taste or smell), even though with a low yield, is obtained using the process developed within the framework of CNR's finalized project [10, 57].

(b) *The whole seed,* using a dehulling process in order to obtain kernels with a very low fiber content. Partial dehulling of the seeds, leaving up to about 10–12% residual hulls, is rather commonly carried out industrially by means of different technologies [9, 17, 23, 40, 52]. The resulting advantages are:

- decreased volume of product to defat, with the subsequent increase of plant capacity;
- decreased wear of the presses;
- better quality of the oil due to a lower wax content;

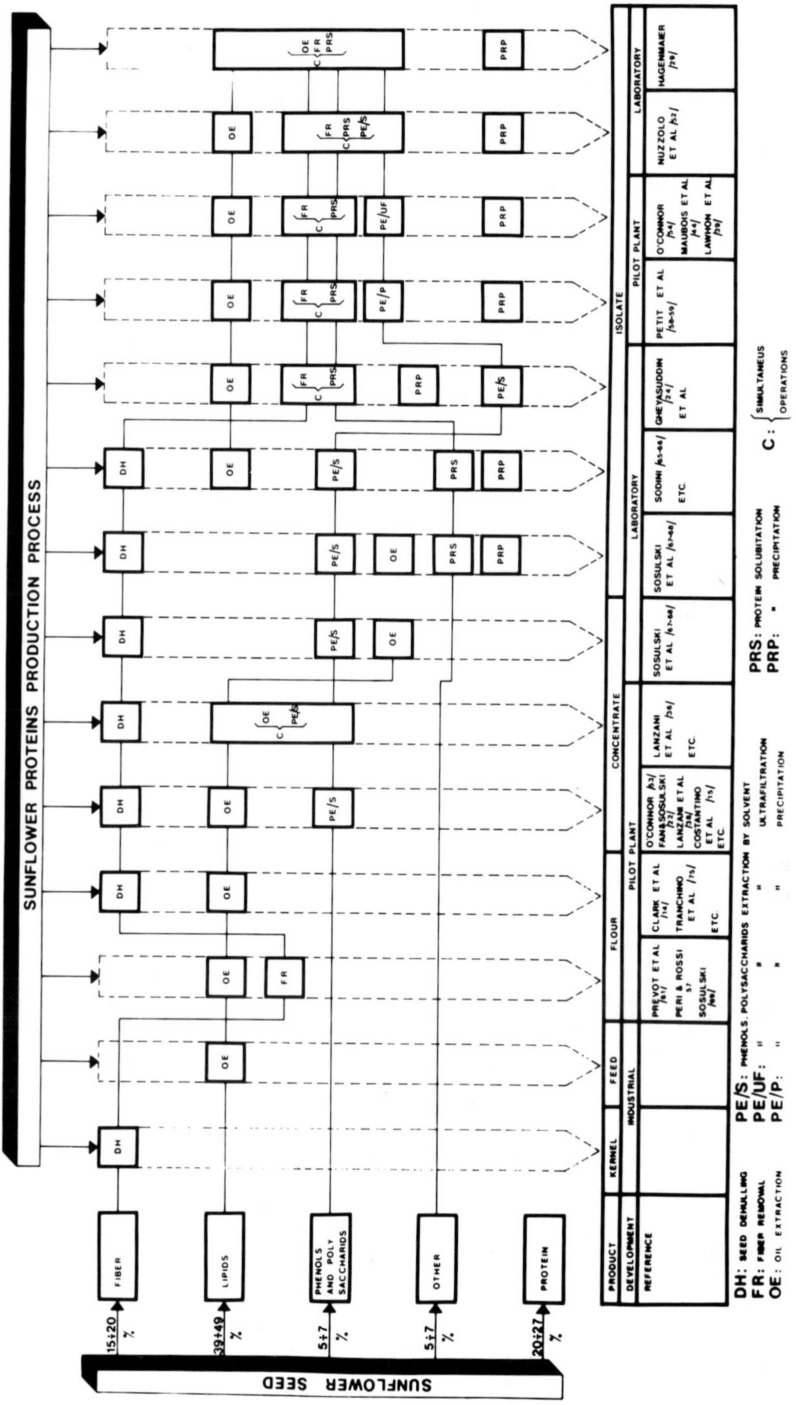
SUNFLOWER PROTEINS PRODUCTION PROCESS
SUNFLOWER SEED
FIBER
LIPIDS
PHENOLS AND POLY SACCHARIDS
OTHER
PROTEIN
15÷20 %
39÷49 %
5÷7 %
5÷7 %
20÷27 %
DH
OE
FR
PE/S
PE/UF
PE/P
PRS
PRP
C
PRODUCT
DEVELOPMENT
REFERENCE
KERNEL
FEED
FLOUR
CONCENTRATE
ISOLATE
INDUSTRIAL
PILOT PLANT
LABORATORY
PREVOT ET AL /61/
PERI & ROSSI 57
SOSULSKI /60/
CLARK ET AL /14/
TRANCHINO ET AL /75/
ETC.
O'CONNOR /53/
FAN & SOSULSKI /22/
LANZANI ET AL /38/
COSTANTINO ET AL /15/
ETC.
LANZANI ET AL /36/
ETC.
SOSULSKI ET AL /67-68/
SOSULSKI ET AL /67-68/
SODINI /65-66/
ETC.
GHEYASUDDIN ET AL /24/
PETIT ET AL /58-59/
O'CONNOR /54/
MAUBOIS ET AL /44/
LAWHON ET AL /39/
NUZZOLO ET AL /52/
HAGENMAIER /29/
DH: SEED DEHULLING
FR: FIBER REMOVAL
OE: OIL EXTRACTION
PE/S: PHENOLS, POLYSACCHARIDS EXTRACTION BY SOLVENT
PE/UF: " " " ULTRAFILTRATION
PE/P: " " " PRECIPITATION
PRS: PROTEIN SOLUBITATION
PRP: " PRECIPITATION
C: SIMULTANEUS OPERATIONS

Figure 1

Table 4. Fractionation of sunflower meals obtained by direct solvent extraction

Meal	Process	Yield	Product
Meal N × 6.25 (dry basis) = 21.3% Ref.: Peri & Rossi [59]	Sieving	21.0%	Protein concentrates N × 6.25 = 59.5% crude fiber = 4.8%
		10.5%	Flour N × 6.25 = 47.1%
		68.5%	Fibrous product N × 6.25 = 5.7%
Meal N × 6.25 (dry basis) = 28.1% Ref.: Prevot et al. [61]	Sieving	29.8%	Flour HTP N × 6.25 = 46.4% Cellulose = 7.7%
		11.5%	Flour MTP N × 6.25 = 45.2%
		18.1%	Flour BTP N × 6.25 = 31.7%
		40.5%	Fibrous product N × 6.25 = 8.6%
Meal N × 6.25 (dry basis) = 37% Ref: Sosulski [69]	Liquid cyclone	50%	Flour N × 6.25 − 56% crude fiber = 7%
		50%	Fibrous product N × 6.25 = 18%

– better quality of the extraction flour due to an increase of the protein content.

An almost complete dehulling (hull residues of about 1–3%) allows for limiting to a minimum the transfer of pigments from the hulls to the flour [14, 80] as well as reducing the fiber content of the finished product which can be then considered suitable for human use. Industrial applications regarding complete seed hull removal are unknown.

The Food Protein Research and Development Center, Texas University (FRDC), carried out detailed research regarding the feasibility of producing hull-free kernels by means of different industrial technologies [42]:

– rubber roll dehuller;
– single disc attrition mill;
– corrugated roll dehuller;
– bar dehuller.

The results of this study, and those of other researchers [37, 60, 63, 64] show the difficulties of dehulling high oil sunflower seeds, but do suggest, however, the adoption of technologies based on impacting the seeds on to a target.

Research mainly devoted to the dehulling of high oil seeds with an air jet impact dehuller have been carried out by ASSORENI [74]. The effectiveness of dehulling has been measured as a function of the characteristics of the seed (variety, moisture, etc.) and of the operating parameters (impact velocity, etc.). The results obtained show the following effects:

- the variety of the seed strongly affects the degree of dehulling. It is well known that 'confectionary' seeds can be dehulled much more easily than 'high oil' seeds [37, 42, 60, 63, 64]. However, even within the high oil seeds, there is a perceptible variation in dehulling effectiveness in relation to variety (Figure 2). In any case it is interesting to note that the more

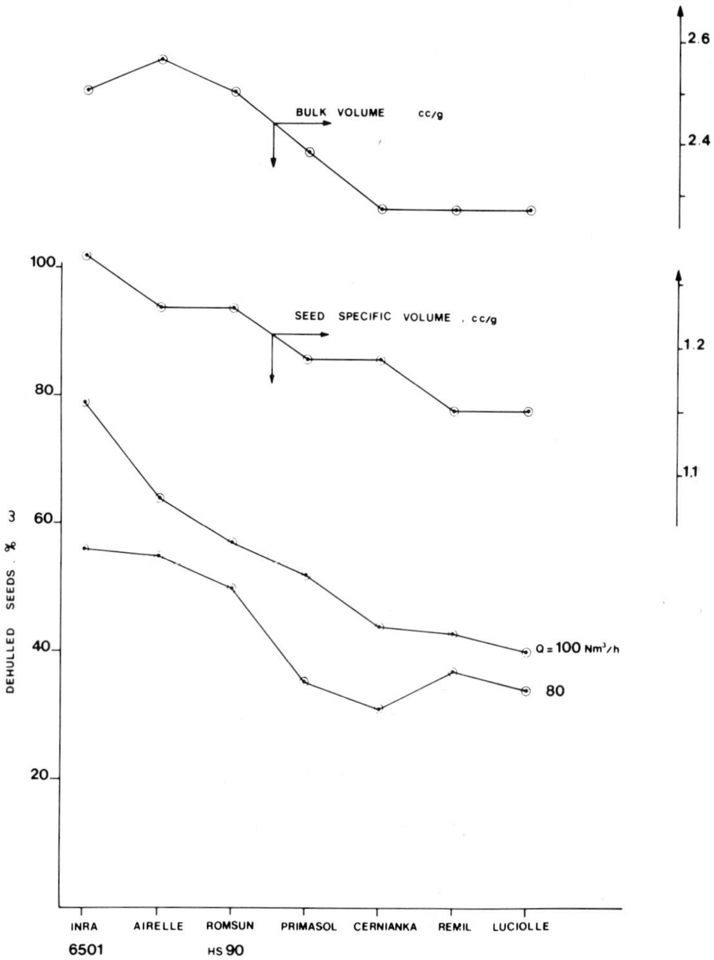

Figure 2. Influence of sunflower seed variety on the percentage of dehulling, and on bulk and seed specific volume.

easily dehulled varieties (Inra, Airelle, etc.) have a lower apparent and/or actual density (measured with the method described in [31]). This observation is understandable in terms of a greater air interspace between the hull and the kernel, and allows for roughly predicting the dehulling characteristics of commercial grade seeds.

- a decrease in the moisture of the seed increases the percentage of dehulled seed (100–(U); Figure 3), but also increases the quantity of product having a dimension less than 2 mm (fines, (F)). Preliminarily, a moisture of about 3% may be considered optimum as it allows reaching a minimum amount for the sum(U) + (F) (Figure 3). The proper optimum, however, is determined by considering the system performances of the integrated dehulling-separation operation and is found to be in the range of 3–5%, depending upon the characteristics of the seeds.
- increasing the impact velocity of the seed on the target, expressed as transport air flow rate in the acceleration tube, increases the quantities of dehulled seeds and fines, simultaneously, with a preliminary optimum at about 100 Nm^3/h for the acceleration tube tested (Figure 4). The performances of the integrated dehulling-separation system may be

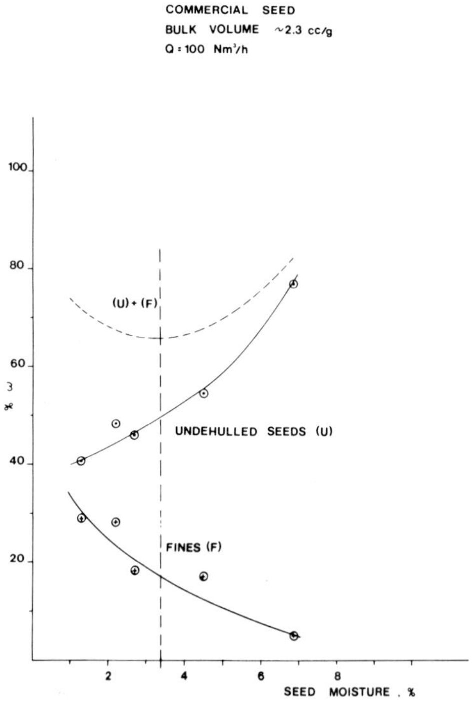

Figure 3. Percentage of undehulled seed and of fines (< 2 mm) as a function of seed moisture.

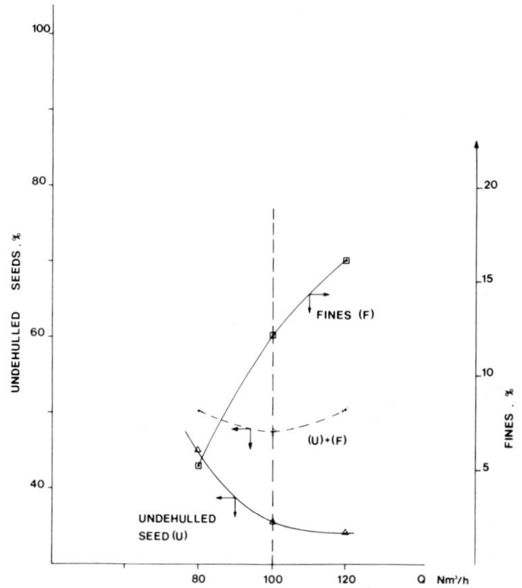

Figure 4. Percentage of undehulled seed and of fines (< 2 mm) as a function of the impact velocity, expressed vs. transport air flow rate.

characterized by curves, relating the kernel recovery vs the hull content in the final product for a given set of parameters related to seed characteristics and operative variables. Figure 5 shows an example of these curves obtained with a Hydromécanique et Frottements plant consisting of an air jet impact huller and a fluidized bed separator [5]. A negligible kernel loss (less than 0.5%) occurs if a high percentage of hulls (greater than 6%) is allowed within the air current of the dehulled product. On the other hand, the production of completely hull-free kernels requires a very high rate of product recycling back to the dehuller with greater fines formation and kernel losses (about 15%) in the hull stream.

In order to have an acceptable product at the completion of extraction, it is necessary to limit the hull content to about 2–3%; therefore in the example shown, there is a kernel yield of 95–97%.

Lipids extraction

The extraction of lipids may be carried out:

(a) directly on whole seeds: these are the conventional oil extraction technologies. The aim is maximum recovery of oil without any attention being given to the quality of the meal which is usually destined for animal feed;
(b) on partially, or nearly totally dehulled seeds: these are processes which

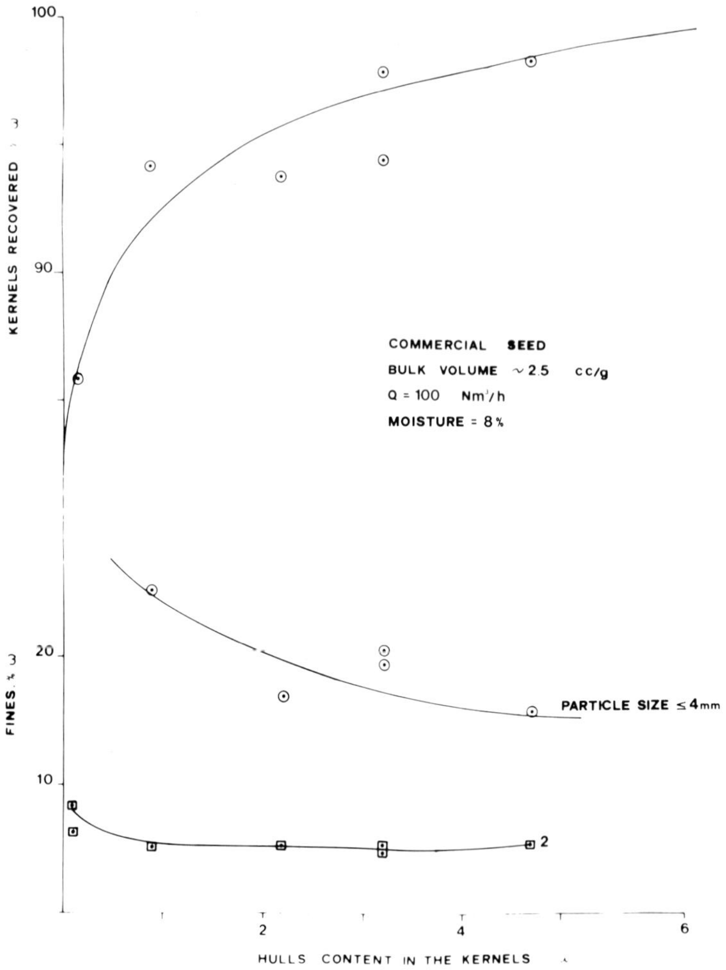

Figure 5. Kernel recovery and fines present in the final product as a function of the hull content.

emphasize, to a greater extent, producing flours (higher protein content and less fiber content) including 'food grade' flours;

(c) on dehulled and dephenolized seeds: these involve those processes which have been used on a laboratory scale by various authors as an alternative to dephenolization after defatting. Oil extraction from dephenolized kernels, however, does not seem to present specific problems [22, 67, 68].

(d) on dehulled seeds with simultaneous extraction of phenols: these processes are aimed at the production of partially defatted products, or concentrates.

Prepressing and solvent extraction processes are the most common procedures used to defat whole seeds and partially dehulled seeds. For economic reasons, prepressing is used to obtain all but some 15–25% of the oil in the kernel; subsequently, extraction with solvent (usually hexane) allows nearly total extraction of the oil (lipid residues between 1–1.5%).

Solvent extraction, either as a subsequent phase to prepressing or a direct operation, is carried out industrially with two types of processes [7, 10, 43]:

Immersion processes. These consist of a solid-liquid contact stage in which the continuous phase is the solvent, with a subsequent stage for separation of phases. This procedure is more suitable for removing the last fraction of oil contained in the seed, which is the most difficult to extract, and allows attainment of low amounts of residual oil in the flours.

Disadvantages of these types of extractors are the low concentrations of oil in the solvent (maximum 15% compared to 25% with percolation plants) and consequently the high costs of solvent recovery. The conventional plants of this type are very bulky and cannot reach high production rates. For these reasons, percolation extractors have been used lately to replace nearly all immersion extractors.

Percolation processes. These are systems which continuously pass a solution through the stationary bulk of seeds. An essential condition for the application of the process is the possibility of draining the solvent through the solid. The procedure is particularly suitable for obtaining the first fraction of oil (about 70%) which is most easily removable. The most serious disadvantages of the percolation extraction process are that it is difficult to obtain good recoveries of the remaining residual oil left in the flour after the first extraction, and above all, it cannot be used on materials which do not permit an easy drainage of the solvent, due to the size and/or physical characteristics of the product being extracted. For this reason, almost completely dehulled sunflower seeds are not easily treated with this type of extractor. Up to now, other types of industrial processes capable of extracting oil from completely dehulled seeds are not known.

Some data regarding oil extraction from dehulled seeds have been given [10, 14, 42]. These indicate the extractive and rheological difficulties in this material treatment.

To overcome these problems a research activity aimed at determining an appropriate extractive system, suitable for treatment of dehulled seeds, was carried out in the framework of the above mentioned ASSORENI research program [56, 75]. The rheological difficulties were bypassed in reconsidering an immersion process, and the extractive problems were solved by means of a proper selection of the operative variables. To realize the effect of these variables, the mechanism of the extraction and the characteristic parameters involved must be taken into account. From the results obtained, the oil diffusion within the solid particles seems to be the rate-controlling step of the process; and the parameters affecting the extractive effectiveness appear to be:

- kinetic parameters: characterizing the oil diffusion rate within the solid particles;
- thermodynamic parameters: characterizing the distribution of lipids between the two phases under equilibrium conditions;
- technological parameters: characterizing the effects of homogenization and retromixing in the systems used. The most important of these is the coefficient of imbibition (i), which determines the quantity of solvent present in the solid after the solid-liquid separation;
- extraction parameters: characterizing the type of contact between solid and liquid (simple or multiple, cocurrent or countercurrent, etc.), the residence time (t), filling ratio (G), etc.

Each of these factors separately affects the extractive effectiveness, even though the main effect comes from their interaction. The equations characteristic of the elementary units of solid-liquid contact that constitute the different industrial extractive systems were derived in order to clarify this interaction. To reach simple explicit correlations, from which the effect of the parameters are made evident, the following simplified hypotheses were used:

- achievement of equilibrium conditions in every stage;
- linearity of the equilibrium relation: $C_L = \alpha C_S$;
- constancy of the imbibition coefficient i.

The correlations derived are shown in Table 5, and are graphically illustrated in Figure 6, in the case of an equilibrium coefficient $\alpha = 1.3$ kg flour/kg hexane, and imbibition coefficient $i = 0.35$ kg hexane/kg flour, corresponding to the system under discussion. The same figure shows the experimental results obtained with the two-stage mixer-settler countercurrent system, and with the five-stage semicontinuous system with a filling coefficient $G = 0.11$ kg flour/(kg flour + kg hexane) (curve 6), and $G = 0.31$ (curve 7). It appears to notably show effects of technological parameters (i) and of extraction parameters (G); the main effect, however is that of the equilibrium coefficient (α) during the initial phase of extraction, and the kinetic coefficient of the transfer of oil (K) in the final phase.

The possibility of simultaneous lipid and polyphenol extraction was studied within the framework of the above mentioned CNR finalized project [36]. The process involves a brief boiling water treatment of dehulled seeds kept in a homogeneous dispersion. Partially defatted meals (51% protein, 22% lipids) and oil have been obtained at both the laboratory and the pilot plant scales. The simultaneous extraction of lipids and polyphenols may be carried out with a hexane-ethyl alcohol-water ternary solvent. It is then

Table 5. Simplified correlation of elementary units of solid-liquid contact

EXTRACTION SYSTEMS	CORRELATION
1. Single stage mixed vessel 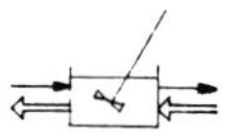	$\frac{(Cs)_i}{(Cso)} = \frac{1 + i\alpha}{1 + R\alpha}$ with $R = \frac{L}{S}$
2. Two stages cross flow 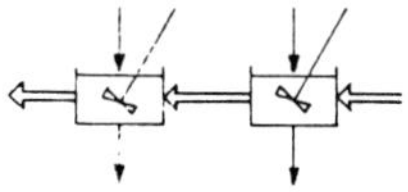	$\frac{(Cs)_i}{Cso} = \frac{(1 + i\alpha)^2}{(1+\alpha R/2)(1+i\alpha+\alpha R/2)}$
3. Two stages countercurrent 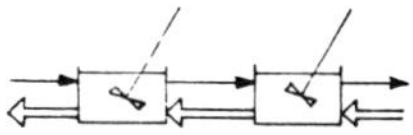	$\frac{(Cs)_i}{Cso} = \frac{(1 + i\alpha)^2}{\alpha^2 R^2 + \alpha R + (1 + i\alpha)}$
4. Single stage pseudo-batch (stationary solids) 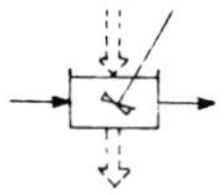	$\frac{(Cs)_i}{Cso} = (1+ i\alpha) e^{-R^{*}\alpha}$ with $R^{*} = \frac{L \cdot t}{S}$, $G=1$
5. Infinite stages pseudo-batch (plug flow liquid, stationary solid) 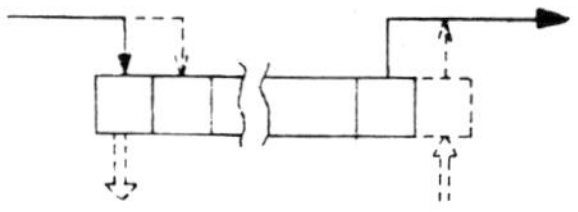	$\frac{(Cs)_i}{Cso} = 1 - \frac{\alpha R}{1+i\alpha}$

possible to obtain easily separable phases, good yields in oil, and limited denaturation of the protein.

Elimination of polyphenolic compounds

Polyphenolic compounds, and in particular chlorogenic acid, are responsible for giving undesirable colorations to sunflower protein products when used as foodstuff, as well as reducing the bioavailability of some essential amino acids. This phenomenon occurs at alkaline pH due to the chemical or enzymatic oxidation of polyphenols to quinones [10, 12, 62]. They react with some functional groups of the protein structures (amino groups of lysine, indole groups of tryptophan, sulfur bridges of methionine) with the formation of chromogenic groups in which the amino acid is bonded in a non-biodisposable form and is nutritionally unavailable.

Various attempts have been made for removing polyphenolic compounds, utilizing different raw materials according to different processes, in order to

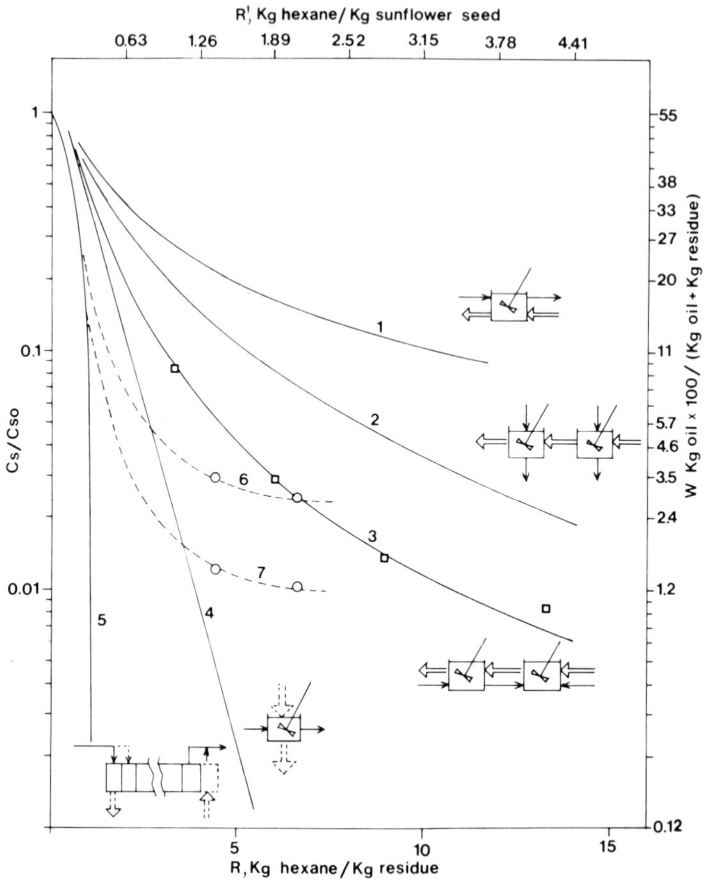

Figure 6. Residual oil concentration as a function of the extraction ratio for different extraction systems, with $\alpha = 1.3$ and $i = 0.35$ [29]: (1) Mixed extractor, single-stage. (2) Mixed extractors, two-stages (cross flow). (3) Mixed extractors, two-stages (countercurrent flow). (4) Semicontinuous mixed extractor, single-stage, $G = 1$. (5) Infinite semicontinuous extractors (countercurrent flow), $G = 1$. (6) and (7). Five semicontinuous extractors, $G = 0.11$ and $G = 0.31$ respectively, experimental results. □ Experimental results of system 3.

obtain protein products with good biological and functional qualities taking into account the technical-economic aspects of production:

(a) Dephenolization of dehulled seeds: the removal of phenols is carried out before or simultaneously with defatting. A very detailed work on the extraction of phenols from dehulled seeds with acidulated water at a pH of 4.5 was carried out by Sosulski et al. [22, 67, 68]. The fundamental advantages of this type of process, compared to those of dephenolization from flour, are basically rheological: easy mixing and separation of the liquid-solid system. From the extractive point of view, this process exhibits some advantages compared to the extraction from flour with water. The seed membrane serves to contain or reduce the diffusion of protein, lipids, etc. (compared to extraction from flour with water, protein loss is reduced from 25–26% to 12–17%. However, it is undoubtedly less efficient than the extraction from flour with ethanol (protein loss ~ 3%; Table 6).

(b) Dephenolization of defatted flour with acidulated water or ethanol: dephenolization with acidulated water presents the above mentioned rheological difficulties as well as problems from the extractive point of view, with relatively low yields, about 60% (cases 3–8 in Table 6); or low protein content, about 66% (case 10 in Table 6); or very high denaturation (case 7 in Table 6) [22, 38, 53]. From the point of view of rheology and extractability, however, the system improves substantially if the dephenolization of the flour is carried out with a water-ethanol solvent. In this solvent, protein extraction losses decreased by about ten times due to the reduced solubility of the proteins. Therefore, the protein content in the final protein product (73–77%) and the yield (77–80%) increase simultaneously (cases 11–13 in Table 6). This line, which appears to be the most promising to obtain a sunflower proteic concentrate, was studied in detail by Assogna et al [4] and Costantino et al [15].

The scaling up of this laboratory and bench-scale data was verified during the operation of the pilot plant. The predicted values of extractive efficiency (cases 12 and 13 of Table 6) were confirmed with a certain scattering which reflected differences in the characteristics of the starting seeds.

Isolation of Proteins (Isolates)

All of the non-proteic components of sunflower seeds or derived products (flour, concentrate) may be removed with a protein solubilization operation at alkaline pH, elimination of the solid residue, and reprecipitation of the proteins when the solution is brought to an acid pH.

This procedure, when applied to seeds or flour that have not been dephenolized, gives rise to the undesirable oxidation of polyphenols (mentioned above in point 4), with the subsequent formation of undesirable coloration and reduction of the bioavailability of some amino acids. For this reason, the isolation of protein is preceded by or combined with dephenolizing operations. In any case, it must be remembered that the characteristics of the raw material are of fundamental importance in determining the efficiency of the dephenolization, the yield in protein, and the characteristics of color, solubility and purity of the isolate.

In particular, according to Sodini et al. [65], the pressing, but above all, the toasting stages in the production of the meal, strongly reduce the solubility of the proteins, hindering the removal of polyphenols and carbohydrates bonded irreversibly to the proteins (Table 7).

The processes to obtain protein isolate may be divided into the following principal classes:

Table 6. Influence of extraction conditions on the yield and characteristics of sunflower

N	Extraction conditions: Procedure	Raw material	solvent	pH	T (C°)	Number of extraction (N)	Total solvent to flour ratio (l/kg)	Total extraction time (min)
1.	Countercurrent Diffusion	Seed	Dilute HCl	4.5	80	5	6	150
2.	Idem	Seed	Dilute HCl	4.5	65	6	6	180
3.	Countercurrent Extraction	Flour	Dilute HCl	4.5	80	5	6	150
4.	Idem	Flour	Dilute HCl	4.5	24	5	6	150
5.	Idem	Flour	Dilute HCl	4.5	24	6	6	180
6.	Cross flow Extraction	Flour	Dilute HCl	4.5	24	5	25	150
7.	Idem	Flour	Dilute HCl	4.5	80	5	25	150
8.	Idem	Flour	Dilute HCl	4.5	60	6	90	108
9.	Idem	Flour	Dilute HCl	5	100	1	25	0.17
10.	Idem	Flour	Dilute HCl	5	100	3	75	0.50
11.	Countercurrent	Flour	Ethanol 70% Vol.		24	5	6	150
12.	Idem	Flour	Ethanol 75% Vol.	4.5	30	2	16	120
13.	Semicontinuous Countercurrent	Flour	Ethanol 75% Vol.	4.5	25	6	7.5	180

Table 7. Influence of technological processes on the extractability of phenols and on isolate protein yield obtained from sunflower meal

Processing		Concentrate		Isolate	
Defatting	Drying	Chlorogenic acid (%)	Total carbohydrates (%)	Protein	Color
Pressing	Toasting	0.9	1.5	46	Dark green
Direct-solvent extraction	Toasting	0.6	1.3	53	Brown
Direct-solvent extraction	Low temperature 30°C	0.3	0.5	80	Creamy

concentrate

Yield		Protein concentrate						
Extract protein loss (%)	Dry weight yield (%)	N × 6.25 (%)	Nitrogen solubility index (%) pH 5	7	9	9.5	Chloro-genic acid (%)	Ref.
12.1	73.8	70.5		9	41		0.4	[22]
17.0	69.6	70.6		13	69		0.8	
26.5	60.0	72.5					0.8	
24.9	60.6	73.4		43	90		1.2	[22]
26.1	57.2	76.5		57	94		0.5	
25.2	58.5	75.7		53	89		0.5	[22]
30.0	60.1	69.0		27	44		0.3	
	53.0							[53, 54]
			15			85	0.4	[38]
20.0	70.0	66.0	23			75	0.1	
2.8	78.0	73.8		18	76		0.4	[22]
3.0	80.0	70.0	9	25	64	66	0.3	[75]
3.0	78.0	73.9				72	0.3	[15]

(a) preliminary removal of polyphenols with the above-mentioned processes (leaching of the seeds or flours with water and/or ethanol) and subsequent isolation of the protein.

These kinds of processes certainly give a good quality product. The sole difficulty is the cost of preliminary double extraction (defatting and dephenolization). Total yields of isolate of 24% and 45%, respectively from flour or concentrate, can be deduced from the data reported by O'Connor [53] on dehulling/defatting/leaching with the acidulated water/protein isolation process. Substantially higher total yields were obtained with regards to dehulling/defatting/leaching using the ethanol/protein isolation process. The protein yields are 70% and 72% from flour and concentrate, respectively. In this case, the isolates have a creamy white color and a protein content (N x 6.25) of about 95% [65, 66].

(b) solubilization of the proteins in a nondephenolized product at an alkaline pH in the presence of an antioxidant and subsequent elimination of polyphenols through precipitation [58, 59] or ultrafiltration [16, 39, 44, 54].

The research work carried out by INRA along both of these lines have contributed to extremely interesting pilot plant results.

In order to avoid the oxidation of polyphenols into quinones (which are more reactive towards proteins), the preliminary phase of solubilization of the proteins is almost always carried out in the presence of an antioxidant (sodium sulfite [16, 24, 44], an inert gas [54], a chelating agent [52], or a surface active agent (sodium dodecil sulfate [39]) which allows solubilization of the proteins at an acid pH, thus hindering the oxidation of the polyphenols.

Nevertheless, these precautions are not generally sufficient to prevent the reaction of the proteins with the formation of chromogenic groups in the subsequent phase of precipitation of the isolate. Therefore, to obtain an isolate with good characteristics it is necessary to carry out subsequent dephenolization after the antioxidant treatment:

(1) Gheyasuddin [24] operates an extraction with isopropanol on the isolate.

(2) O'Connor [54], Culioli and Maubois [16, 44] and Lawhon [39] suggest an ultrafiltration treatment on the protein solution. Commercial microporous membranes with an approximate pore diameter of between 15 and 70 A and usable, and they make a separation at the molecular weight level of around 10,000–100,000. The low molecular weight polyphenol components are thus separated from the proteins. The isolate is obtained by direct atomization or dispersion or spinning of the protein solution, or, alternatively, by precipitaion at acid pH, washing and drying. The protein content of the isolate is in the range 87–97% and the protein yield is 40–60%, depending on the procedure used. The technical-economic feasibility of this technique will be dependent upon obtaining high permeation rates and membranes with a long technical life.

Another approach for eliminating polyphenolic constituents was suggested by Petit [58, 59]. The procedure consists of the precipitation of colloids and the pigments of the hulls with calcium acetate, and that of phenols with magnesium acetate, during the solubilization of the proteins in an alkaline medium. The protein isolates, obtained after the isoelectric precipitation (pH 4.5–5.5), or atomization or spinning of the protein dispersion, have a nitrogen content of 16–16.5% and are completely soluble in an alkaline medium.

The use of chelating agents, such as the aluminate ion reported by Nuzzolo [52], allows for the formation of stable complexes at alkaline pH between polyphenols and AlO_2^- or $Al(OH)_4^-$; the oxidation of the polyphenols is thus then avoided during the course of the solubilization of the proteins. The subsequent acidification of the solution by means of citric acid allows obtaining a protein precipitate at the isoelectric point; simultaneously citric acid, acting as a complexing agent, maintains the aluminium in solution. A creamy white isolate with 95% protein content and 0.4% polyphenol is obtained with a yield of about 60%.

Finally, a completely different technique was suggested by Hagenmaier [29] which permits treating the whole seed directly with an alkaline solution for the solubilization of proteins in the presence of Na_2SO_3 and for the simultaneous extraction of oil. The proteins are then reprecipitated from the aqueous solution. However, rather serious difficulties are encountered in the solid-liquid separation and in the breaking of the water/oil emulsion. Further, the yield of oil is low and the isolate is of a dark color.

Solvent stripping

The problem of removing solvent from vegetable protein products is extremely important from the point of view of health, and with respect to the flavor of the products themselves.

Only few data on the levels of residual solvents in sunflower products are present in the literature: 300–900 ppm of hexane in edible flour prepared by the FPRDC [14], and about 200 ppm of hexane in the toasted flakes from the food oil industry [40]. On the other hand, much data are available for soybeans, although they are rather conflicting: the makers of desolventizer guarantee only a negative POP TEST; in a recent review Chessin [13] reported data showing 55 to 225 ppm of hexane in commercial soybean edible flours; Honig et al. [30] reported data indicating 90 to 410 ppm of hexane in commercial soy protein products; they also show a large variation in the effectiveness of various drying procedures for removing solvents from hexane-ethanol azeotropic extracted soy flours (Table 8).

All of these data are presented without an interpretation of the phenomenon, and therefore generalizations or extrapolations are not possible. But if the phenomenon is carefully studied, the following main effects appear:

- during the drying of protein products, an approximately linear reduction of the water content and an asymptotic reduction of the solvent content occurs as a function of the time (Figure 7);
- the value of the asymptote changes with varied drying conditions; more precisely, the faster the heating of the flour, the higher the asymptote.

These facts allow for interpreting the process in terms of the simultaneous transfer of heat and matter in which:

(a) the rate of water removal is controlled by the transfer of heat of the saturated evaporating surface;
(b) the rate of solvent removal is controlled by the rate of diffusion of the solvent within the flour particle.

On this basis, it is possible to formulate a mathematical model of the

Table 8. Levels of volatiles ($\mu g/g$) in azeotrope-extracted soy flours after various desolventizing procedures (Honig et al. [30])

Desolventizing procedures	Volatile component					
	Methanol	Ethanal	Ethanol	Acetone	Hexane	Hexanal
1. Air dried	3	3	3.600	11	2.200	4
2. Freeze dried, twice at 0.7 mm Hg after wetting to 35% moisture (w35%)	34		124		350	
3. Vacuum strip, no steam, on rotary evaporator at 50°C, w 35%	5		32		130	2
4. Vacuum strip, steam, on rotary evaporator whithe blending in steam	7	0.1	23	4	470	5
5. Toasted, steam heated at 100°C for 10 min	5		19	16	12	3

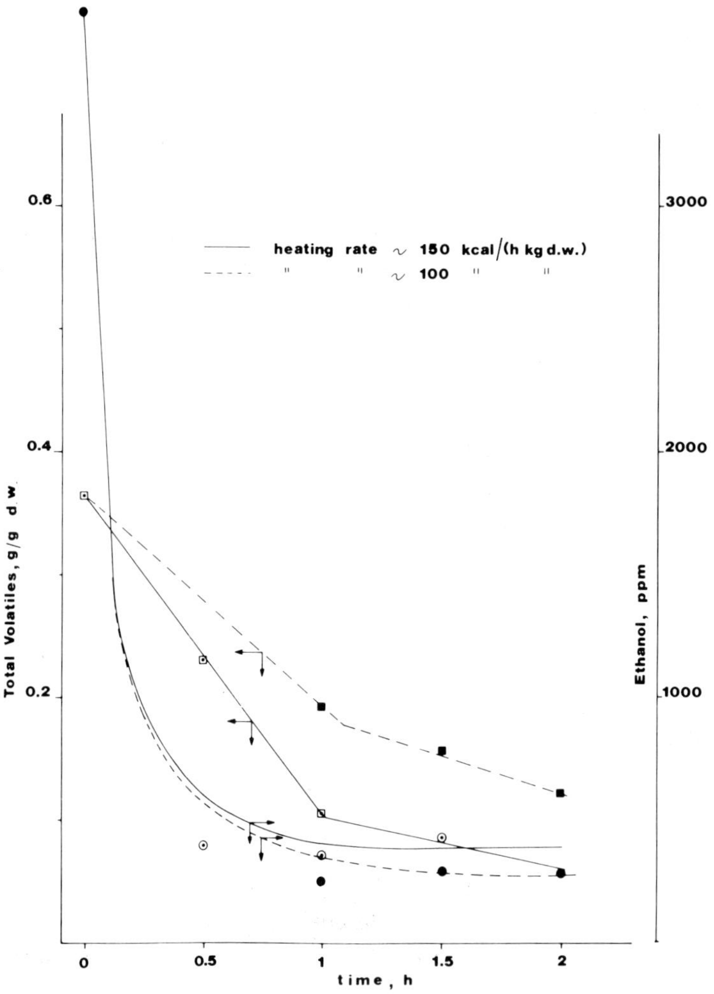

Figure 7. Water and Ethanol content of sunflower concentrate as a function of the time for the two heating rate values.

process which allows understanding the above-mentioned effects (theoretical curves in Figure 7). The same analytical approach appears satisfactory for interpreting both the removal of hexane from flour and of ethanol from protein concentrate, except for the different expression of the diffusion coefficients. Sufficiently low amounts of residual solvent in the proteic products are achievable if low heating-rate, wetting, and grinding of the products are adopted.

Microbiological control

Great emphasis is given in the literature to the necessity of microbiological control in the plants used for production of food grade soybean proteins [47, 50]. However, for soybean, the seed thermal conditioning and the toasting or steam treatment of the meal allows for having low aerobic total counts (about 10^4 germs/g dry weight) with no problem.

For sunflower there are less data, but increases from 10^3 to 10^5 in the total count with the presence of coli have been reported during the finishing of food grade sunflower seed flour [14].

A statistical study concerning the microbiological content of commercial sunflower seeds showed that the most probable total count is around 10^6 with the presence of entero bacteria and E. coli (Figure 8 [18]).

The effect of processing on the microbiological content of sunflower products, as measured in the continuous pilot plant, is indicated in Figure 9, for the case of an oil extraction with hexane at 40°C, and of a phenol extraction with a 70% ethanolic solution at ambient temperature without thermal conditioning of the seeds or final toasting of the products. A notable reduction in the total microbiological count is observed only in the treatment with the water-alcohol solution. In any case, successive increases have been noted in the count during the finishing of the product. However, the protein concentrate obtained, contained a total count of about 10^5, without E. coli and pathogens; results which are better than observed in good commercial practice.

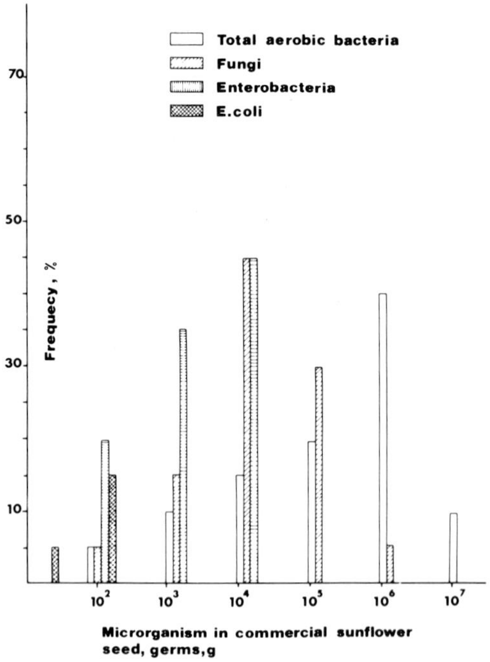

Figure 8. Statistical distribution of microrganisms in commercial sunflower seed.

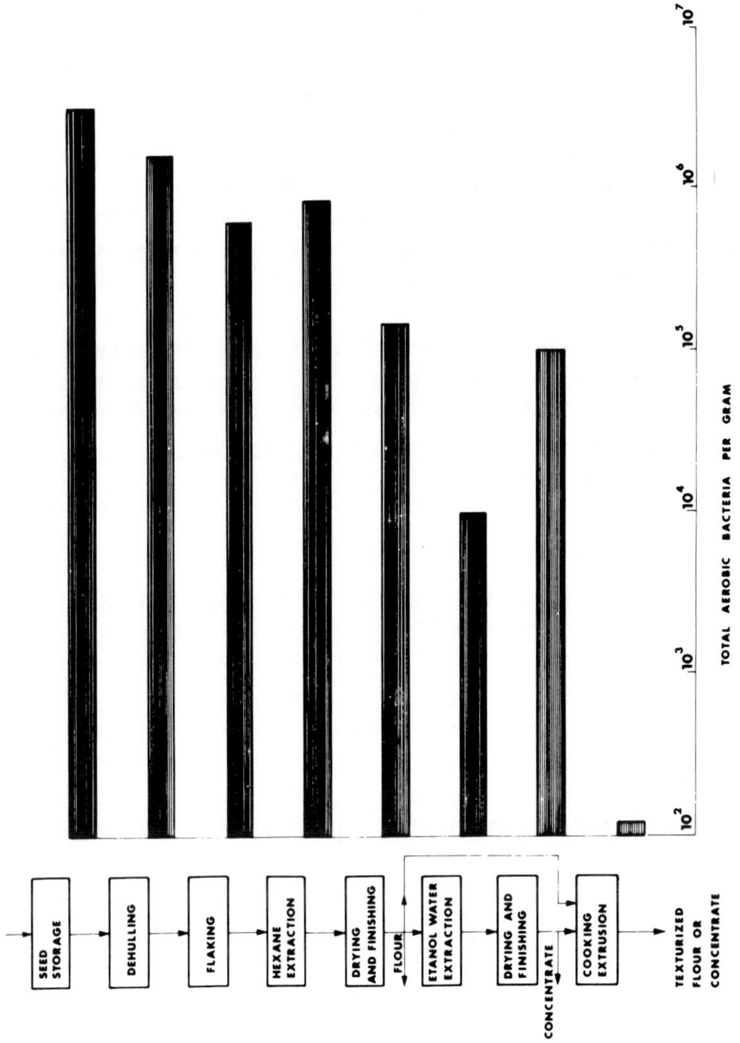

Figure 9. Effect of processing on microorganisms in sunflower products.

In the case of flour production, to obtain the same result, it is necessary to include a sanitizing phase in the process which avoids the transfer of a possible contaminant of the seed to the product. This phase may be a thermal conditioning of the seed or a final toasting of the flour.

The production of texturized protein by cooking-extrusion appears to be very efficient in reducing the total aerobic count to very low values (Figure 9).

Rapeseed

As with many oil seeds, rapeseed contains some undesirable factors. With rapeseed, erucic acid is present in the lipid fraction and glucosinolates are detectable in the flour. Both of these factors have hindered the diffusion of rapeseed and the utilization of its products for food. Considerable research was devoted to overcoming these problems, working in two main directions:

Genetic improvement: attempting to select varieties with low contents of low erucic acid and thioglucosides ('double zero' types). This research has shown good results with respect to the reduction of erucic acid (from about 50% of the total content of the lipids to about 2% in the new selected varieties). Meanwhile, further work is required for the reduction of the thioglucosides content which is present in the range 3–7% in the most common varieties.

Detoxication of the flour by removal of thioglucosides. Really, these substances, biologically inactive by themselves, are hydrolyzed by the enzyme myrosinase, contained in the seed as well, to produce toxins and bitter substances (nitriles, isothiocyanates etc.), which negatively affect the thyroid, the kidneys, the liver and other organs more or less seriously [3, 49, 81].

Actually, industrial rapeseed extraction meal is used as ruminant feed because the hydrolysis products of the thioglucosides are destroyed in the rumen, the only negative factor being in the nasty flavour. The use of rapeseed proteic products for monogastric animals, or human consumption, however, requires the removal of thioglucosides and/or the inactivation of the myrosinase enzyme.

The problems inherent to the preparation of raw material, the eventual dehulling, the oil extraction, and the proteic enrichment of the rapeseed meal, are similar to those already examined for sunflower [3, 5]. So only the problems specific to rapeseed processing will be discussed, i.e., myrosinase inactivation, thioglucosides extraction and protein isolation.

Myrosinase inactivation and glucosinolates removal

A preliminary thermal conditioning of the seeds (80–90°C for about 15 minutes) for myrosinase inactivation is always included in current rapeseed processing to improve both meal and oil quality. Afterwards, the extracted meal is toasted to eliminate the volatile compounds formed by hydrolysis of thioglucosides. The thermal treatment of the seeds is not sufficient to detoxify rapeseed meal because even though myrosinase is completely inactivated, it is not possible to exclude possible hydrolysis of the thioglucosides through other nonenzymatic methods and, in any case, the possibility exists that myrosinase can be absorbed with other foods. So meal detoxication is effected only by removing, almost completely, the thioglucosides. For this the following processes can be adopted:

The extraction processes seem to be the most promising ones to prepare proteins for human consumption. These processes are characterized by the solvent (water, ethanolic solution or water-acetone mixture), the product to be extracted (whole or ground seed or meal), and by the processing system used. According to Belzile and Bell [6] and Solsulski et al. [8], the detoxication of rapeseed proteic products is more effective if the hydrolysis of the more water soluble thioglucosides is avoided.

The processes suggested by these authors therefore, include inactivation of myrosinase (heating at 100°C for 3 min [2, 6, 8], or treatment with an 80%(vol.) ethanol solution [3]) and then extraction of the thioglucosides with water, ethanolic solutions or water-acetone.

On the contrary, other authors [73, 76, 77, 78, 79] suggest the hydrolysis of the thioglucosides (enzymatic and/or thermic) before extraction with water or water acetone.

The processes pursuing the first idea are the most numerous. Of those we mention the Karlshamns patented process [1, 55] for the production of food grade rapeseed protein concentrate by water extraction of dehulled and ground seeds; the combination of the production line of food grade products with that of conventional oil production (pressing) allows building up an integrated system in which the by-products of the first line (fines, hulls, extract) find a valuable use, while also avoiding pollution problems.

Eapen et al. [18] suggest a similar process starting with enzyme inactivation (by cooking the seed in water for 2 min) and then removal of glucosinolates by acqueous extraction.

A detailed analysis of the acqueous extraction of thioglucosides from ground seed as a function of pH has been carried out by Staron and Guillaumin [70]. This analysis resulted in the determination that pH 8 was the pH value at which the yield of dry matter (68%), protein (70%) and oil (95.6%) were maximum. So a patented process for the production of an edible, high quality oil and of a food grade protein concentrate with high yields was developed [71].

Bhatty and Sosulski [8] suggest, on the contrary, a treatment of the whole rapeseed with a 50% (by vol.) ethanol solution (pH 12 with NaOH). An efficient diffusion of glucosinolates with minimum losses of other seed constituents (proteins and lipids) occurs. The ethanol both inhibits myrosinase and makes thioglucoside dissolution easier. The treated seed may be suitable, after partial dehulling and air classification, for protein concentrate preparation.

Also, Eklund et al. [20] suggest the use of an ethanolic solution (about 30% vol.) with NaCl to extract the thioglucosides from dehulled seeds. The direct use of the defatted product (48% protein, 31% lipid, 0.4 mg/g thioglucosides) was also proposed.

Protein isolate (isolates)

Rapeseed protein isolate can be prepared with about the same processing as for sunflower isolate. But the presence in rapeseed of proteins with widely different characteristics (molecular weights from 10,000 to 300,000 [3]; isoelectrical points from pH 3.6 to 11), would require, for obtaining good proteic yields, multistage solubilization and precipitation. Kodagoda et al. [33] obtain three distinct isolates with yields respectively of 11; 13 and 42%, with isoelectrical precipitation of proteic extracts obtained with water, with HCl at pH 2, and with NaOH at pH 10. The presence of precipitation coadjutants can nevertheless allow obtaining better yields of isolates and/or limiting the number of solubilization and precipitation stages.

In fact, as shown in [25], it is possible to increase the yields in dry matter (from 16 to 33%) and in nitrogen (from 35 to 75%) by adding sodium phytate to the rapeseed proteic extracts. Unfortunately the phytic acid is likely to exert an adverse effect on the nutritional value of the isolate so that other types of precipitation coadjutants have been studied: hexametaphosphate (HMP), carboxymethyl cellulose (CMC), electrolytes etc. [26]. An almost quantitative recovery of the extracted proteins was achieved with each one of the following schemes [41]:

- precipitation at pH 6.6 (Isolates A) followed by precipitations at pH 5 in presence of CMC (Isolate B). The isolate A is rich in high molecular weight proteins with isoelectrical points from 4 to 7; the isolate B is rich in middle-low molecular weight proteins with isoelectric points from 6 to 8.
- precipitation at pH 4.9 in the presence HMP (Isolate C), formed by proteins of various molecular weights with isoelectrical points from 4.5 to 10.

The presence of HMP, in the protein extraction phase also appears to increase the yield of protein isolated. An increase in nitrogen extraction from 61 to 81% was obtained when the HMP concentration was brought from 0 to 2% [72].

Greater protein yields (about 90%) are reported by Nockrashy et al. [21] following a counter-current protein extraction with acqueous solution of Na OH (0.02 N) and double precipitation at pH 6 and pH 3.6.

Yield, however, depends very much on the characteristics of the raw material (denaturation during myrosinase inactivation, etc.) and on the protein losses in the solid-liquid separation and in the isolate washings. The yield of solubilized nitrogen in NaOH (0.1 N) decreases from 89% to 62% if the rapeseed meal prepared under mild conditions is replaced by a rapeseed meal coming from oil industries [27].

The glucosinolate content in the isolate is usually much lower than in the starting meal. Thompson [72] reports that more than 98% of the glucosinolate initially present in the meal (12.11 mg/g) was removed during protein extraction, isolate precipitation, and washing. Therefore, the total glucosinolate level obtained in the isolate (0.14 mg/g) was not considered to involve any negative nutritional effect. However, the total exclusion of the thioglucosides and other chromogen impurities from the isolate is possible by means of ultrafiltration of the protein extract as suggested for sunflower isolate production [44].

Conclusions

In the present discussion, the problems related to the various processes suggested by different authors for producing edible sunflower and rapeseed protein products, have been briefly examined.

Some of these processes are in an advanced stage of development and can be used to produce economically feasible yields of good quality products for food manufacturing industries. Nevertheless, to date, no food grade sunflower or rapeseed protein is found on the market. However, sunflower and rapeseed protein sources will have to be taken into consideration in the very near future as a promising and potentially useful means for attacking food shortage problems.

References

1. Anjou K, Fecske A (1974) The production of Rapeseed Protein Concentrate, Proc. 4th Int. Rapeseed Conf. Giessen, 4–8 June 1974, 659–662
2. Appelqvist L-Ä, Josefsson E (1967) Methods for quantitative determination of isothiocyanates and oxazodinethiones in digests of seed meals of rape and turnip rate. J Sci Food Agric 18:510–519
3. Appelqvist L-Ä, Ohlson R (1972) Rapeseed. Elsevier Amsterdam-London-New York, 101–121
4. Assogna A, Patricelli A, Sodini G, Emmi E (1979) Factors affecting the extraction of chlorogenic acid from defatted sunflower meal. Lebensm Wiss u Technol 12: 262–265
5. Baudet M (1981) Le depelliculage des graines de colza. Bulletin Cetiom 79:16–23
6. Bell JM, Belzile RJ (1965) Rapeseed meal for livestock and poultry: a Review. The Canada Department of Agriculture, 45
7. Bernardini M (1971) Tecnologia Olearia, Tecnologie srl, Roma
8. Bhatty RS, Sosulski FW (1972) Diffusion extraction of rapeseed glucosinolates with ethanolic sodium hydroxide. J Am Oil Chemists' Soc 49:346–350
9. Brueske G (1980) Sunflowers. Proc ISF/AOCS World Congress, New York
10. Cantarelli C (1981), Il Girasole, una fonte di proteine alimentari. CNR, Milano Tip S Teresa, 5–105
11. Carboner A (1950) Comptes Rendus de l'Academie d'Agriculture de France, no 14
12. Carter JF (1978) Sunflower science and technology. American Soc. of Agronomy, Madison, USA, 407–440
13. Chessin RL (1981) Investigating solvent residuals in soybean meal and solvent content of vented streams. Proc 72ns AOCS Meeting, New Orleans
14. Clark SP, Wan PJ, Matlock SW (1980) Pilot plant production of sunflower seed flour. J AOCS, 57:275A–279A

15. Costantino R, Assogna A, Patricelli A, Sodini G (1981) Produzione di concentrato proteico da farina di girasole. Proc Progress in Food Eng, Milano
16. Culioli J, Maubois JL (1975), Purification des protéines de tourteau die tournesol par ultrafiltration. Rev Franc Corps Gras, 10
17. Defromont C (1972) Decorticage due tourneol. Proc 5th Int Sunflower Conf Clermont-Ferrand, 353–361
18. Di Maggio D, Marghinotti D, Assogna A, Patricelli A (1981) Microbiological and biochemical aspects of Italian sunflower cultivars. Proc of FAO Consultation on the European Cooperative Network on Sunflower, Pisa
19. Eapen KE, Tape NW, Sims RPA (1969) New process for the production of better quality rapeseed oil and meal. II: Detoxification and dehulling of rapeseeds. Feasibility study. J Am Oil Chemists' Soc 46:52–55
20. Eklund A, Agren A, Langler T (1971) Rapeseed protein fractions, I. Preparation of a detoxified lipid-protein concentrate from rapeseed by a water-ethanol extraction method. J Sci Food Agr 22:650
21. El Nockrashy AS, Mukherjee KD, Mangold HK (1977) Rapeseed protein isolated by countercurrent extraction and isoelectric precipitation. J Agric Food Chem 25(1): 193
22. Fan TY, Sosulski FW, Hamon NW (1976) New techniques for preparation of improved sunflower protein concentrates. Cereal Chem 53:118–125
23. Fetzer W (1980) Sunflower dehulling today. Proc ISF/AOCS World Congress, New York
24. Gheyasuddin S, Cater CM, Mattil KF (1970) Preparation of colorless sunflower protein isolate. Food Tech 24:36–37
25. Gillberg L, Törnell B (1976) Preparation of rapeseed protein isolates J Food Sci 41:1063
26. Gillberg L, Törnell B (1976) Preparation of rapeseed protein isolates. J Food Sci, 41:1070
27. Girault A (1973) The study of some properties of rapeseed protein with a view to protein concentrate production. J Sci Food Agric 24:509–518
28. Goering KJ, Thomas OO, Beardsley DR, Curran WAJ Jr (1960) Nutritional value of mustard and rapeseed meals as protein source for rats. J Nutr 72:210–216
29. Hagenmaier RD (1974) Aqueous processing of full fat sunflower seeds: yields of oil and protein. J AOCS, 51:470–471
30. Honig DH, Warner KA, Selke E, Rackis JJ (1979) Effects of residual solvents and storage on flavor of hexane/ethanol Azeotrope extracted soy products. J Agric Food Chem 27:383–1386
31. Jones JD (1943) Intergranular spaces in some stored foods. Food 12:325–328
32. Kirk LD, Mustakas GC, Griffin EL Jr (1966) Crambe seed processing: improved feed meal by ammoniation. J Amer Oil Chem Soc 43:550–555
33. Kodagoda LP, Yeung CY, Nakai S, Powrie WD (1973) Preparation of protein isolates from rapeseed flour. J Inst Can Sci Technol Aliment 6(3):135–141
34. Lanzani A, Jacini G (1973) Attuali conoscenze sui composti solforati ad azione gozzigena presenti nel seme di colza. Riv Ital Sost Grasse 50:277
35. Lanzani A, Petrini MC, Jacini G (1974) Le sostanze ad azione antitiroidea contenute nel seme di colza nel quadro della produzione di isolati proteici. Riv Ital Sost Grasse 6:113
36. Lanzani A, Cardillo M, Petrini MC (1979) Tecnologia integrale di estrazione e preparazione di farine a partire da semi di girasole. Riv It Sost Grasse 56:107–110
37. Lanzani A, Petrini MC (1979) Un impianto per la sgusciatura del seme di girasole. Riv It Sost Grasse 56:204–205
38. Lanzani A, Cardillo M, Petrini MC (1979) Preparazione di concentrati proteici da semi di girasole per via umida. Riv It Sost Grasse 56:48–51
39. Lawson JT, Rhee KC, Lucas EW (1981) Soy protein ingredients prepared by new processes. Aqueous processing and industrial membrane isolation. J AOCS, 377–384
40. Leibovitz Z, Ruckenstein C (1980) Modern processing of sunflower seed. Proc ISF/AOCS World Congress, New York

41. Lönnerdal B, Gillberg L, Törnell B (1977) Preparation of rapeseed protein isolates: a study of rapeseed protein isolates by molecular sieve chromatography. J Food Sci 42:75
42. Lucas EW (1978) Final report on development of technologies for the production of low fiber sunflower seed meal and food grade flour, Texas University
43. Martinenghi GB (1963) Tecnologia chimica Industriale degli oli, grassi e derivati, Hoepli, Milano
44. Maubois JJ, Culioli J, Chopin A, Chopin MC (1976) Ultrafiltration process for obtaining protein isolates of vegetable origin, US Patent 3, 993, 636
45. Mielke S (1981) Oil World Statistics, ISTA MIELKE & C, Hamburg
46. Miller RW, Van Etten CN, McGrew C, Wolff IA, Jones O (1962) Amino acid composition of seed meals from forty-one species of cruciferae. J Agr Food Chem 10:426–430
47. Milligan ED, Amlie JH, Reyes J, Garcia A, Meyer B (1981) Processing for production of edible soya flour. J ACOS, 58:331–333
48. Mossé J, Baudet J (1972) Biochemical aspects of quality and content of lysine in sunflower seed proteins. Proc 5th Int Sunflower Conf, Clermont Ferrand, 437–440
49. Mustakas GC, Kirk LD, Griffin EL Jr, Clanton DC (1969) Crambe seed processing. Improved feed meal by soda ash treatment. J Amer Oil Chem Soc 45:53–57
50. Mustakas GC, Milligan ED, Taborga JA, Fellers DA (1980) Conversion of soybean extraction plant in Bolivia to production of flours for human consumption: feasibility study. J AOCS 57:55–58
51. Neumunz GM (1980) Sunflower seed dehulling systems. Proc. ISF/AOCS World Congress, New York
52. Nuzzolo C, Vignola R, Groggia A (1980) Method for preparing a proteinic isolate from sunflower seed meal using aluminium salts. US Patent 4, 212, 799
53. O'Connor DE (1971) US Patent 3, 586, 662. In: Hanson LP (ed) Vegetable protein processing. Noyes Data Corporation, London 1974
54. O'Connor DE (1971) US Patent 3, 622, 556. In: Hanson LP (ed) Vegetable protein processing. Noyes Data Corporation, London 1974
55. Ohlson R, Anjou K (1979) Rapeseed protein products. J Am Oil Chemists' Soc 56:431–437
56. Patricelli A, Assogna A, Casalaina A, Emmi E, Sodini G (1979) Fattori che influenzano l'estrazione dei lipidi da semi decordicati di girasole. Riv It Sost Grasse 56:136–142
57. Peri C, Rossi M (1980) Produzione di farine proteiche di girasole. Tecnologia di preparazione e qualità delle proteine. Riv It Sost Grasse 57:465–469
58. Petit L, Davin A, Gueguen J (1974) Procédé pour l'obtention d'isolates purifies de proteines, Brevet d'Invention 74, 35594
59. Petit L, Davin A, Gueguen J, Lefebvre (1975) Extraction et purification des proteines par voie chimique: tourteau de tournesol et farine de fève. Rev Franc des Corps Gras 10:517–520
60. Popova DL, Serdyuk VI, Kopeykousky VM (1968) Contribution to the question of reasons for the problems on high oil sunflower seed dehulling. Masloz Prom 34: 7–12
61. Prevot A, Bloch C, Defromont C (1972) Enrichissement en protéines des tourteaux de tournesol par voie mecanique. Proc 5th Int Sunflower Conf, Clermont-Ferraud, 368–384
62. Robertson JA, Burns EE (1975) Use of sunflower seed in food products. Critical Reviews in Food Science and Nutrition, 201–240
63. Sallans HR, Sinclair GD (1945) A laboratory huller for sunflower seed and oats. Can J Research, F 23:306–312
64. Sinclair GD, Sallans HR (1947) Flash drying in relation to hulling of sunflower seed and oats. Can J Research F 25:119–127
65. Sodini G, Bernardi A, Bittoni A, Canella M (1976) Influenza dei trattamenti industriali sulla estraibilità di genoli e carboidrati da farina di girasole e relativa solubilità delle proteine. Riv Sci Tecn Alim Nutr Um 219–221
66. Sodini G, Canella M (1978) Methods for extracting phenols and oligosaccharides from vegetable tissues. US Patent 4, 072, 671

67. Sosulski FW, McCleary CW, Soliman FS (1972) Diffusion extraction of chlorogenic acid from sunflower kernels. Journal of Food Science 37:253–256
68. Sosulski FW, Sabir MA, Fleming SE (1973) Continuous diffusion of chlorogenic acid from sunflower kernels. Journal of Food Science, 38:468–470
69. Sosulski FW (1981) Fractionation of oil seed meals into flour and hull components with liquid cyclones. Proc Progress in Food Eng, Milano
70. Staron TJ, Guillaumin R (1979) Une méthode d'éxtraction des protéines et de l'huile de colza par l'eau. Rev Fran Corps Gras 11:441
71. Staron TJ (1979) Brevet d'Invention INRA V F no 2417263, 14
72. Thompson LU, Allum-Poon P, Procope C (1976) Isolation of rapeseed protein using sodium hexametaphosphate. J Inst Can Sci Technol Aliment 9:15
73. Tookey HL, Van Etten CH, Peters JE, Wolff IA (1965) Evaluation of Enzyme-modified, solvent-extracted Crambe-seed meal by chemical analyses and rat feeding. Cereal Chem 42:507
74. Tranchino L, Melle F, Sodini G (1982) Almost complete dehulling of high oil sunflower seeds to be presented at 73rd AOCS Meeting, Toronto
75. Tranchino L, Sasselli G, Melle F, Grillo G (1981) Oil extraction from dehulled sunflower seeds. Proc Progress in Food Eng, Milano
76. Vaccarino C (1975) Detossificazione della farina di colza. I: Trattamento idrolizzante del seme macinato prima dell'estrazione dell'olio, Riv It Sost Grasse 52:91–94
77. Vaccarino C (1975) Detossificazione del seme di colza. Nota II: Trattamento idrolizzante sul seme intero. Riv It Sost Grasse 52:195–197
78. Vaccarino C, Toscano MA, Tripodo MM (1976) Detossificazione del seme di colza mediante un trattamento idrolizzante sul seme intero. Riv It Sost Grasse 53:291–294
79. Vaccarino C, Chiofalo L, Tripodo MM, Toscano C (1978) Dctossificazione del seme di colza mediante un trattamento idrolizzante sul seme intero. Riv Ital Sost Grasse 4:271–276
80. Wan PJ, Baker GW, Clark SP, Matlock SW (1979) Characteristics of sunflower seed and meal. Cereal Chemistry, 56:352–355
81. Youngs CG, Perlin AS (1967) Fe(II)-catalyzed decomposition of sinigrin and related thioglycosides. Can J Chem 45:1801–1804

The proteins from leaves

ROBERTO FIORENTINI and CARLO GALOPPINI

Istituto di Industrie Agrarie dell'Università, Via S. Michele degli Scalzi, 4, 56100 Pisa, Italy

Key words: leaf protein concentrate, production processes, uses, human food

Abstract. The nature and main characteristics of the leaf proteins are reviewed, stressing the even aminoacid composition of the different crops suitable for protein extraction. A satisfactory pattern of essential aminoacids is also shown to be a general characteristic.

The principles of the wet fractionation of green vegetables are then reviewed; details are given on the squeezing stage and the criteria for the recovery of the proteins from the expressed juice are reported. The different physico-chemical status of such proteins requires different processes for their recovery as a whole or in fractionated bulk. Therefore, three different leaf protein concentrates (LPCs) can be obtained: cytoplasmatic (white), chloroplastic (green) and whole LPC, the last being a combination of the other two.

Processes for LPC production as animal feed have already been established at the industrial level; procedures giving LPC for human consumption are at the laboratory or pilot scale. In this paper, attention has been paid to the criteria for the preparation of edible LPC, which are essentially based on both the decoloration of green whole LPC and the fractionated recovery of the cytoplasmatic and chloroplastic proteins present in the juice. In this second process, the white fraction is reserved for humans and the green LPC for animals.

Technological aspects of the production of the above-mentioned LPCs are described and some nutritional and functional properties of the final products are briefly illustrated.

Introduction

One of the most serious problems presently facing mankind is the shortage of protein, most of which is of vegetable origin.

Since it is highly unlikely that enough protein can be obtained from animal sources to meet the increasing demand due to the growth in population, increasing the value of vegetable proteins becomes of great importance. There are three main ways of doing this:

(a) increase the total production of food of vegetable origin;

(b) genetically improve the quality of the proteins obtained from the crops;

(c) make a more rational use of the vegetable proteins which are at present available.

We shall limit ourselves in this discussion to examining the third point, with particular reference to the possibility of using leaf proteins in the human diet. It should be pointed out here that there is a vast amount of literature on this topic, with more than 1000 bibliographical references up until now. Most of this work has been carried out in the last 30 years [10, 31, 42]. The amount of scientific work dedicated to this sector is justified when one considers that leaves represent the world's largest source of protein and that the production of protein per hectare for some of the more noted forage crops is much greater than that for the seeds of the most important cereals and legumes.

Leaf proteins

The chemical composition of the leaves, unlike that of the seeds, is generally fairly uniform, especially with regards to the protein fraction which is, on average, 10–20% of the dry matter [36].

Almost three-quarters of the proteins in the epigean part of green crops are concentrated in the cells of the mesophyll. These cells are characterized by a large central vacuole and a reduced cytoplasmatic layer, where almost all of the cellular proteins are contained. Around 80% of these proteins are concentrated in the chloroplasts where they can be divided almost equally between the soluble phase (stroma) and the system of lamellar membranes (thylakoids). The majority of the remaining 20% makes up the cytoplasmatic gel, while there are relatively small amounts in the nucleus (1–2%) and mitochondria (< 5%) [36].

On the basis of its water solubility, leaf protein material can be divided, almost equally, into insoluble and soluble proteins. The insoluble fraction consists mainly of the lamellar proteins of the chloroplasts, where they are associated with chlorophyll, carotenoids and lipids [21]. One protein, commonly known as Fraction-1, accounts for almost half of the soluble fraction. The location of this protein is the stroma, where it carries out an enzymic activity essential to the fixation of the CO_2 during photosynthesis. The Fraction-1 protein, though chloroplastic of origin, is not associated with pigments or lipids. It has been repeatedly isolated in the leaf extracts of numerous plants and in the case of tobacco it has also been obtained in crystalline form [33, 52]. It corresponds to the enzyme ribulose diphosphate carboxylase and has a molecular weight of just over 500,000 daltons. There is now considerable information available on Fraction-1, which being present in fairly large quantities in all photosynthetic tissues, is one of the most abundant proteins in the world. In the remaining soluble protein material, there is a large amount of free aminoacids and oligopeptides which reflect the high level of protein synthesis taking place in the mesophyll [21, 36].

As far as the aminoacid composition of the leaf proteins is concerned, it is worth remembering that the sequence of aminoacids in the proteins is genetically coded and that such a sequence, in the homologous proteins, can only undergo slight modifications and then only on condition that the substitution of one or more aminoacids does not significantly alter the biological properties of the protein. For this reason, the photosynthesis and other biological activities in the leaves require that the sequence of aminoacids in the proteins is rigidly followed. The research carried out by Gerloff et al. [26] confirms this; after analyzing the aminoacid composition of the proteins extracted from the green part of various crops, they found that in every case the variations were minimal, even for different conditions of fertilization and harvest times.

The aminoacid composition of leaf proteins will be given further on in this discussion. For now it is sufficient to say that it can be profitably compared to that of proteins of animal origin. This reveals that leaf proteins, contrary to the traditional idea that they can only be of use in the feeding of ruminants, have a high nutritional value even for monogastrics, including man.

Because of their uniform composition, all leaf tissues can be considered as a potential source for the production of protein concentrate, although as far as protein yield per hectare is concerned some species are more productive than others as a result of their greater luxuriance.

The protein yield, however, does not just depend on the characteristics of the plant, since there are numerous other agronomic and environmental factors which can have decisive effects. The harvesting is extremely critical, since the bringing forward or delaying of the ideal time, even by just a few days, can cause a considerable drop in protein yield per hectare. The optimum harvesting time for the extraction of the leaf proteins for most plants coincides with the pre-flowering stage, although in some cases it can extend into the more or less advanced stage of flowering [3, 41].

The use of fertilizers also has an important role in determing the yield of protein per hectare. Of the most commonly used fertilizing agents, it is nitrogen that has the greatest effect, helping to produce more luxuriant plants. It produces the best results in the ammonia form, whose activity is even more evident if used with nitrification inhibitors [41].

Irrigation has also shown itself to be particularly effective in increasing both the unit yield in protein and its extractability. This is because the vegetable tissues are more turgid and, therefore, richer in protein juice [40].

In the light of what has been described above, it appears evident that the ideal crop for the production of leaf protein concentrates is a plant which is luxuriant and highly efficient in photosynthesis, which reacts well to nitrogen fertilizers or uses atmospheric nitrogen, and which rapidly synthesizes the proteins and lignifies slowly. It should also contain only small quantities of acids, tannins, and mucilages and no toxic substances. There are in practice numerous plants, both wild and cultivated, that have such properties, while others could be developed genetically.

The protein yield of some of the most common crops used in artificial meadows and pure hayfields are shown in Table 1. All the crops mentioned in this table have been shown to be suitable for the preparation of leaf protein concentrates. Numerous other non-forage crops, as well as the green by-products of industrial vegetables, aquatic plants and even tree leaves have also been successfully processed [41].

Table 1. Protein yield of selected forage crops

Forage crops	Protein yield (kg/ha)
Leguminosae:	
Medicago sativa	2800
Trifolium pratense	3000
Trifolium repens	2200
Onobrychis viciaefolia	2200
Hedysarium coronarium	1500
Vicia faba minor	1250
Trifolium alessandrinum	2200
Trifolium incarnatum	1100
Trigonella foenum graecum	1000
Vicia spp	1100
Vigna sinensis	1100
Graminaceae:	
Avena sativa	800
Hordeum vulgare	1100
Secale cereale	800
Zea mays	1800
Sorghum vulgare	1700
Cruciferae:	
Brassica napus	1250
Brassica rapa	1000

Wet fractionation of grasses

The extraction of leaf proteins for the production of protein concentrates (LCPs) for use in animal feeds or in the human diet has been made possible by the technology of the wet fractionation of grasses.

The wet green crop fractionation basically consists of a mechanical operation of grinding and pressing which enables the fresh forage to be separated into two fractions: a protein juice and a press cake. The degree of squeezing (extraction) has a considerable effect on the characteristics of the juice and press cake. Three different degrees of extraction are recognizable: a mild extraction, a partial extraction and an exhaustive extraction.

The mild extraction is used in conjunction with green crop drying and attempts to remove the maximum amount of water commensurate with the minimum loss to the juice of dry material. A heat treatment of the forages prior to pressing reduces the extractability of the leaf proteins, which are then more easily retained in the leaf tissues. In the case of the mild extraction, the juice is not worth processing because of its low dry matter content and low protein level. It is only useful, therefore, as a fertilizer or poor protein integrator in feeds for monogastrics.

The partial extraction is based on the fact that many crops, such as alfalfa, have a protein level which is generally greater than that required in the feeding of ruminants. This extraction provides a press cake, which is better suited to the feeding of ruminants, and a leaf juice, containing the extracted protein surplus, from which LPCs for animal (monogastrics) and human consumption can be profitably produced.

The aim of the exhaustive extraction is to recover as much of the protein as possible in the juice. Here, the fibrous press cake is of secondary importance and primary attention is devoted to the preparation of protein products from the juice. The full extraction includes a vigorous pressing which is usually carried out twice and preceded by suitable treatments such as grinding the forages finely, alkalinization, etc. The press cake from this extraction is high in fibre, poor in protein and, therefore, of low nutritional value. It can be used as a feed for ruminants after nitrogen fortification, and for monogastrics after aerobic bioconversion [23]. Other foreseeable uses include the preparation of biogas, methanol and ethanol by means of fermentation or physico-chemical conversion [34, 35].

The press juice obtained from the partial or exhaustive extraction has a high level of dry matter, usually more than 10%. The protein content (30–40% of the dry matter) is high enough to enable the preparation of LPCs for animal or human consumption. A number of processes are available for this purpose, ranging from the preparation of whole dehydrated juices to the obtaining of pure proteins in crystalline form. Most of these processes, however, include the mechanical separation of the proteins in the press juice, following their coagulation or insolubilization.

The proteins in the plant juices are made up of two quite different fractions. The first of these, called chloroplastic, consists of insoluble lipoproteins which can be easily destabilized and, once isolated, have a dark green color and a strong grassy flavour. The second fraction partially originates in the cytoplasm and for this reason is called cytoplasmatic. It consists of relatively stable soluble proteins, which when recovered are tasteless, odourless and white or cream in color.

The fractionated recovery of the two fractions can be achieved by exploiting their different physico-chemical properties. The green fraction (chloroplastic LPC) is suitable for feeding monogastrics with an efficient protein conversion (pigs, rabbits, poultry), while the white fraction (cytoplasmatic LPC) can be used directly in the human diet. Alternatively, the two protein fractions can be recovered together thereby obtaining a whole LPC. This is used in animal feeds or can be consumed directly by man, after being first purified or extracted to remove the pigments and other undesirable substances.

Although there are several different methods for the recovery of the juice proteins (heat coagulation, acidification, addition of polyelectrolytes or polar solvents, ultrafiltration, anaerobic fermentation, etc.), the processes which have been developed industrially use heat coagulation by means of steam injection. The subsequent separation of the protein cake from the deproteinized juice is usually carried out by centrifugation or filtration, while the drying is carried out on a fluidized bed drier or by an analogous method.

The deproteinized serum, which is left behind after the protein separation,

Table 2. The main European plants for the wet green crop fractionation

Company	Country	Year of installation	Operating capacity (g grass/h)	Annual production of LPC (t)
Liciencia	Hungary	1973	80	1200
France–Luzerne	France	1976	400	1500
BOCM–Silcock	Great Britain	1977	90	500
Aproalfa	Spain	1978	130	1200
Anhydro	Denmark	1979	440	1600
France–Luzerne	France	1979	800	3000
France–Luzerne	France	1980	800	3000

NOTE: There are no large industrial wet fractionation plants in Italy. There are, however, some pilot plants (operating capacity 30–50 q of grass/h) and others are being developed.

Table 3. Industrial production of whole LPC for animal feeding: yields and chemical composition of the product

Yields:	
– Initial solid transferred to the LPC	7–10%
– Initial nitrogen transferred to the LPC	20–25%
Chemical composition of LPC (dry basis):	
– Protein (N × 6.25)	55%
– Fat	9%
– Fibre	2%
– Ash	11%
– N-free extract	23%
– Carotene	500 ppm
– Xanthophyll	1000 ppm
Essential aminoacids (g/16 g N)	
– Isoleucine	4.9
– Leucine	8.2
– Lysine	6.2
– Methionine + cystine	3.3
– Threonine	4.8
– Tryptophan	1.4
– Tyrosine + phenylalanine	9.3
– Valine	6.2

is rich in sugars and has numerous uses, from a simple fertilizer to a culture medium for producing biomasses. Other uses are the preparation of molasses and the production of biogas and ethanol by fermentation. Although fermentation is the most profitable way of exploiting this liquid by-product of the wet green crop fractionation, the most commonly used technique, on an industrial scale, is its concentration. A molasses-like product is obtained, which is of considerable value for animal feeding because of its high caloric value and good flavour.

The technology of wet green crop fractionation has been developed industrially in several countries, particularly in Europe [22, 23]. All of these processes, however, only produce whole LPCs for use in animal feeds. The technology is fairly standardized involving the heat coagulation at 85–90°C

Table 4. Composition of leaf protein products for human consumption (dry basis) [8]

	Purified whole LPC	Decolorized whole LPC	Cytoplasmatic LPC
Protein (N × 6.25)	55–65	62–74	80–95
Lipid	15–20	2–5	0.5–1.0
Ash	2–5	2–5	0.5–1.0
Fibre	0.5–1.5	0.5–1.5	–
N-free extract	10–25	15–30	5–20

of the proteins in the press juice of alfalfa, which is by far the most commonly used plant. In Table 2 the most important features of the main industrial processes are listed, while in Table 3 the average process yields and the chemical composition of the protein concentrate are shown.

Preparation of leaf proteins for human consumption

In the last thirty years numerous processes for the production of edible leaf proteins have been studied, both in the laboratory and on a pilot scale. These processes can be divided as follows:

– processes for the preparation of purified whole LPCs;
– processes for the preparation of decolorized whole LPCs
– processes for the preparation of cytoplasmatic LPCs.

The average chemical composition of the three products is shown in Table 4. The protein content increases as one passes from the purified whole LPC to the cytoplasmatic LPC. Other differences are linked to the recovery process and will be discussed later.

Purified whole LPCs

The process to obtain purified whole LPCs and the chemical composition of the products, do not differ significantly from those of the feed-grade LPCs. There are certain additional operations in the process when compared to the conventional wet fractionation technology. These operations (washing of the fresh grass, clarification of the juice and washing of the wet protein cake) ensure that the final product is hygienically and otherwise acceptable. Once washed, the wet protein cake can be used directly in foods, preserved by the addition of suitable substances, or dried. Good results in preserving the wet cake have been obtained with 2% acetic acid and 15% lactic acid or NaCl [2, 46].

A dry product, obtained using the process described above, has been fed in India and Nigeria to young children suffering from malnutrition. The results obtained have been consistently good [1, 9, 19, 38, 45, 51]. Whole purified LPCs are the only products that have been widely tested in humans.

Despite their high nutritional value and low production costs, purified LPCs have a number of drawbacks which severely limit their acceptability, especially among western consumers who are notoriously very demanding. These drawbacks mainly concern the dark green color and strong grassy flavour. The grassy flavour can be reduced, but not totally eliminated, by careful selection of the input grass and rapid processing. The colour can also be improved slightly by the addition of reducing agents, such as sodium metabisulphite; however, the results obtained have not been particularly good.

Decolorized whole LPCs

The drawbacks of the purified whole LPCs can be partially overcome by extraction with solvents.

The color of the LPCs is primarily due to the plant pigments (chlorophyll and carotenoids) and, secondarily, to tannic substances (formed by the oxidation of the polyphenols) and the products of the Maîllard reaction. All of these substances can be removed to a large extent by means of suitable solvents which can also considerably lessen the grassy taste of the leaf proteins. The origin of this taste is difficult to define, but it is linked to the presence of specific lipid compounds, such as those formed by the oxidative degradation of the unsaturated fatty acids.

The efficiency of the different solvents in improving the acceptability (color and flavour) of the whole LPCs has been the subject of numerous researches. One of the most important findings has been the impossibility to obtain a good decoloration using dried LPCs; for this reason the extraction is carried out on the wet protein cake [12, 29].

According to Bray [8], four or five countercurrent extraction stages are usually enough to guarantee an acceptable decoloration, provided that the solvent/dry matter ratio is about 10/1 and the temperature is close to the boiling point of the solvent. After extraction, the protein concentrate is dried to a solvent residue level of around 25 ppm, while the solvent is distilled to separate the pigments and remove the water.

Numerous polar, non-polar and mixed solvents have been tested [12, 29, 44]. The non-polar solvents were ineffective, often worsening the color of the LPC, while the polar solvents generally provided good results. This is probably because the wet cake still contains a considerable amount of water (normally more than 50%) and the polar solvents, unlike the non-polar solvents and mixtures, have the capacity to simultaneously remove the phenolic compounds, which are chiefly responsible for the darkening in the color of the products during air-drying. Of the polar solvents tested, acetone, ethanol, propan-2-ol and butanol provided the best results. After further tests, propan-2-ol and butanol emerged as the most promising solvents for extraction on an industrial scale. This is because of their ability to carry out extractions efficiently at room temperature and to be easily recycled.

The extraction with solvents also has the beneficial effect of increasing the protein level of the final product, as a result of the reduction in the lipid content (Table 4). The lipid extraction also increases the time that the product can be preserved. Some authors also suggest that there is an increase in the nutritional value of the decolorized LPCs, but there is not enough literature on this topic to arrive at a definitive conclusion [14, 15, 47]. On the other hand, the extraction with solvents denatures the proteins (already considerably damaged by heat coagulation) and further destroys their functionality.

Although decolorized whole LPCs show a good acceptability, they have not been produced industrially, primarily because of the high operational costs. Even simple solvents, such as acetone, become too expensive in relation to the final product, which is still, after all, fairly poor and not sufficiently profitable. The loss of solvent (around 0.5 kg per kg of decolorized LPC) and the high fuel requirement for the distillation and recovery of the solvent itself are serious disadvantages. It has been suggested that the sale of valuable components (chlorophyll, xanthophyll and carotene), extracted during the decoloration, could make the process sufficiently remunerative, but, in truth, numerous doubts remain as to the economic viability of the technology.

Cytoplasmatic LPCs

The cytoplasmatic LPCs are high in protein, white-cream in color and almost tasteless and odourless. They are obtained by the fractionated recovery of the chloroplastic and cytoplasmatic proteins in the squeeze juice of the grasses.

The chloroplastic fraction in suspension can be easily agglomerated and then removed by centrifugation and/or filtration. It can also be removed directly by means of ultracentrifugation. Since there are no economical machines capable of generating such high centrifugal forces (~ 100,000 × g), the chloroplastic material should be destabilized to enable its removal. This destabilization can be achieved by heating to around 60°C, or by the addition of polyelectrolytes or organic solvents. After the chloroplastic material has been removed, the soluble cytoplasmatic protein can be recovered from the supernatant liquid in several ways (acid or heat coagulation, ultrafiltration, etc.). The non-protein solubles are then removed by washing and the final LPC is obtained by drying on a fluidized bed or by a similar process. Usually a solution of sodium metabisulphite is added to the fresh grass, prior to pressing, to improve the nutritional value of the white protein, the process yield, and the color of the final product.

Numerous processes have been studied and proposed, both at the laboratory and the pilot plant scale, for the preparation of cytoplasmatic LPCs for human consumption. Since the literature available in this topic is already vast, and growing all the time, only the most important processes have been discussed below, with just a brief description of the other processes.

Pro-Xan II process. Among the different processes for the fractionated recovery of the chloroplastic and cytoplasmatic proteins in the plant juices, the Pro-Xan II process deserves a special mention. This process was developed at the W.R.R.C. (USDA Western Regional Research Center) at Berkeley, California, and has been operated on a pilot plant scale since 1975 [20].

In the Pro-Xan II process, the chloroplastic material is agglomerated by heating to 60–65°C for 20 seconds and then isolated by centrifugation and filtration. The soluble cytoplasmatic proteins are isolated from the supernatant liquid by heat precipitation at 80°C followed by centrifugation. The edible protein concentrate (called Welpro from White Edible Leaf Protein) is first washed to remove any non-protein solubles and then dried on a fluidized bed.

The material balance reveals that for alfalfa with about 20% protein (dry basis), the yield of Welpro is approximately a quarter of all recovered LPC, which amounts to about 15% of the initial dry material. The protein content of this cytoplasmatic LPC is 88.7% on a dry basis.

Poly-Protein process. Since heat coagulation denatures the proteins which thus lose useful functional properties, such as nitrogen solubility, studies have been under way for some time at the Istituto di Industrie Agrarie at the University of Pisa, to try and carry out the coagulation of the proteins in the plant juices by using polyelectrolytes instead of the conventional heat denaturation. A large series of polyelectrolytes with high molecular weights and ionic charges of different nature and quantity have been experimentalized. The coagulating activity of these polyelectrolytes takes place at room temperature and so there is no need for any preliminary heating. The results obtained in the laboratory and on a pilot plant scale led to the setting-up, at the beginning of 1976, of a process for the preparation of a whole feed-grade LPC [25].

After setting-up the above-mentioned process, the Institute has been directing its activities towards a new process for the fractionated recovery of the proteins in the plant juices. The aim is to produce both a chloroplastic feed-grade LPC and a cytoplasmatic protein for human consumption. On the basis of the favourable results obtained in the laboratory, a pilot plant was set up in 1979 [24].

The process, called Poly-Protein, is characterized by the coagulation of the chloroplastic fraction by means of a synthetic anionic polyelectrolyte (Superfloc A 150, Cyanamid Italia, Milano), which is added in quantities of about 20 ppm. The cytoplasmatic fraction in the supernatant is recovered by acidification to the isoelectric pH and centrifugation. With regards to the acid coagulation, it should be pointed out that by reducing the pH from 4.5 to a minimum of 3.5, the protein recovered increases to a maximum of 90% of the protein in the juice. By lowering the pH, however, other substances are also precipitated thereby reducing the percentage protein content of the LPC. In the Poly-Protein process, the acidification is carried out at pH 3.5 to maximise the process yield and this justifies the relatively low level of protein in the final product (73.9%).

Since the whole process takes place at room temperature, the proteins are not seriously denatured and show a high nitrogen solubility [24].

The material balance shows that the total yield in protein concentrate (chloroplastic LPC + cytoplasmatic LPC) is about 10% of the initial dry matter, while the amount of nitrogen recovered in these two fractions is around a quarter of the total nitrogen in the fresh grass [24].

Superfloc A 150, a polyacrylamide-based synthetic polyelectrolyte is non-toxic and, as a result of its very low monomer content, is already used in several food technologies [16, 53]. Moreover, it is used in extremely small doses (20 ppm) and remains entirely in the chloroplastic feed-grade LPC.

Coagulation by means of polyelectrolytes is beginning to interest other authors [13, 30]. Much work has still to be done, but the prospects of efficiently producing absolutely safe, undenatured products are good.

Ultrafiltration. The ultrafiltration technique can be used to isolate the cytoplasmatic proteins in a soluble form from plant juices devoid of the chloroplastic fraction [17, 18, 32, 39, 48]. A cytoplasmatic LPC with a maximum protein level of around 60% (dry basis) can be obtained by applying this technique to alfalfa juices.

This protein concentration can only be increased by diafiltration or washing ultrafiltration, which consists of continuously adding water to remove non-protein components by dialysis. The diafiltered LPC has a high concentration of proteins (> 90%) which are in a native state and, therefore, completely soluble, even at pH 6. The high solubility of the diafiltered proteins in acid solutions should also be pointed out; this enables the leaf proteins to be used in slightly acid foods and beverages.

Other processes. A process to prepare cytoplasmatic LPCs with a high solubility, using plant juices devoid of the chloroplastic proteins, has been set up at the W.R.R.C. [17, 18, 37]. This process includes acid coagulation at low temperature and the washing and resuspension in alkali of the cytoplasmatic protein cake. The whole process is carried out at around 2°C and the final product, which contains almost 70% protein, can easily be redissolved in water, at room temperature, even after several months.

The gel filtration also permits the recovery of the cytoplasmatic proteins in the juices devoid of the chloroplastic fraction [17, 18]. Although the yields from this technique are high and the functional properties of the final product are good, it is also very expensive. This means that gel filtration, suitable for the purification of soluble proteins obtained by acid precipitation or ultrafiltration, is only of use when the impurities present considerably affect the functional properties of the final product.

Another process, described briefly by Hove and Bailey [28] and taken up by Bray and Humphries [11], uses organic solvents to destabilize the chloroplastic material in the juice, which is then removed by centrifugation. The

Table 5. Aminoacid composition of alfalfa LPCs (g/16 g N) [7]

Aminoacid	Whole LPC	Cytoplasmatic LPC	FAO's adult aminoacid scoring pattern
Alanine	6.07	6.35	
(Ammonia)	1.61	1.07	
Arginine	6.17	7.98	
Aspartic ac.	10.44	10.04	
Glutamic ac.	11.18	12.02	
Glycine	5.21	5.52	
Histidine	2.45	3.19	
Isoleucine	5.24	5.28	4.0
Leucine	8.92	9.42	7.0
Lysine	6.31	6.43	5.5
Methionine + cystine	3.27	4.33	3.5
Phenylalanine + tyrosine	10.36	11.99	6.0
Proline	4.69	4.44	
Serine	4.42	3.65	
Threonine	5.01	5.70	4.0
Tryptophan	1.65	2.45	1.0
Valine	6.41	6.91	5.0

cytoplasmatic food-grade protein is then recovered by acidification or heat coagulation. A number of solvents were tested: non-polar solvents were efficient at a concentration of 3–4% and polar solvents at a concentration of 10–12%. Solvents with a high specific weight proved to be totally inefficient.

In the Hove–Hobman process, all the proteins in the plant juices are coagulated using polyphosphates, after acidification to pH 2.5. The edible cytoplasmatic fraction is then redissolved at pH 10 and separated by centrifugation and/or filtration from the chloroplastic insoluble fraction [27].

Finally, the France–Luzerne company has set up a process in which the input grass is pre-heated to 60–65°C, so that the chloroplastic proteins are coagulated in the leaf cells. The aim is to obtain a squeeze juice which only contains the edible cytoplasmatic proteins, which can then be recovered by heat coagulation at 80–85°C.

Nutritional value and functional properties of the LPCs for human consumption

The aminoacid composition is one of the most commonly used criteria to judge the nutritional value of a protein. In Table 5, the aminoacid compositions of some food-grade LPCs are shown and compared to the FAO scoring pattern. From this data, it can be seen that the sulphur aminoacids, lysine, threonine, and tryptophan, which are the aminoacids most commonly lacking in the diet, are present in much higher quantities in the cytoplasmatic LPC than in the whole LPC. This is partly due to the addition of sodium

Table 6. Comparison of aminoacid score and E/T ratio of alfalfa LPCs and common foods [7]

Protein	Aminoacid score	E/T ratio
FAO recommendation	100	2.25
Hen's egg	100	3.20
Beef muscle	100	2.78
Cow's milk	95	2.95
Casein	91	3.21
Soybean seed	74	2.46
Rice	66	2.39
Wheat	53	2.05
Maize	49	2.51
Alfalfa LPC (whole)	99	3.02
Alfalfa LPC (cytoplasmatic)	100	3.28

metabisulphite, normally used in the preparation of the cytoplasmatic LPC, which tends to increase the general level of the sulphur aminoacids, by reducing the oxidation of the cystine and methionine during the process [32]. Since these two are the main limiting aminoacids, the PER of the cytoplasmatic LPC can be considerably increased by the addition of metabisulphite. Indeed, when correctly prepared, it has a nutritional value equal to that of casein [32].

In Table 6 proteins of different origin, including two LPCs from alfalfa and the FAO pattern, are compared on the basis of aminoacid score and the E/T ratio [7]. The cytoplasmatic LPC has an aminoacid score as high as 100, the same as beef muscle and hen's egg. Cow's milk, casein, soyabean seed and cereals all have a lower aminoacid score. As far as the E/T ratio is concerned, only eggs have values similar to that of the cytoplasmatic LPC.

As for digestibility, studies carried out in vivo and in vitro reveal that for whole LPC the values are around 80–90%, while for cytoplasmatic LPC, values are usually greater than 95% [7, 43].

Conclusions

All of these data, together with the vast amount of research carried out to study the influence that agronomical and technological factors have on the nutritional value of LPCs, lead us to conclude that leaf proteins, when correctly prepared, have a high biological value and that good proteins, suitable for human consumption, can be obtained from several crops.

The functional properties also play an important role in determining the marketability of the LPCs for human consumption. The functionality of the protein is often held in greater consideration than the nutritional value itself and there are several examples of highly nutritional proteins finding scarce use in the human diet because of their poor functional qualities.

The main functional properties of the LPCs for human consumption have been investigated by several authors, whose works should be consulted for a detailed analysis [4, 5, 6, 49, 50]. Here, it is sufficient to say that LPCs in general and the cytoplasmatic fraction in particular, when correctly isolated, reveal good functional properties, especially with regards to their water and lipid absorption capacity. This makes these LPCs particularly useful as protein integrators in meat and meat-like products. The functionality of the LPCs for human consumption, however, requires more research, especially to establish the influence of the processing conditions.

In conclusion, it should be remembered that the incorporation of whole LPCs in food has been studied by several authors using a vast range of products (vegetable soups, sausages, sauces, biscuits, etc.) with 3–10% LPC [5]. The cytoplasmatic LPCs have also been included in certain foods such as biscuits and vegetable soups; the acceptability of the final products was good at levels up to 3% for biscuits and 15% for vegetable soups [5,6].

References

1. Anonymous (1975) Large scale leaf protein test with Indian children. Food Techn in Australia 27:521–527
2. Arkoll DB (1973) Preservation and storage of leaf protein preparation. J Sci Fd Agric 24:437–445
3. Arkoll DB, Festenstein GN (1971) A preliminary study of the agronomic factors affecting the yields of extractable leaf protein. J Sci Fd Agric 22:49–56
4. Betschart AA (1974) Nitrogen solubility of alfalfa protein concentrate as influenced by various factors. J Food Sci 39:1110–1114
5. Betschart AA (1977) The incorporation of leaf protein concentrates and isolates in human diets. In: Wilkins RJ (ed), Green Crop Fractionation (Proc occasional simp no 9). British Grassland Society, pp 87–95
6. Betschart AA, Kohler GO (1975) Certain functional properties of alfalfa protein isolates prepared and dried by various methods. Proc 12th Techn Alfalfa Conference, Berkeley, ARS-USDA, pp 115–122
7. Bickoff EM, Booth AN, De Fremery D, Edwards RH, Knuckles BE, Miller RE, Saunders RM, Kohler GO (1975) Nutritional evaluation of alfalfa leaf protein concentrate. In: Friedman M (ed), Nutritional Quality of Foods and Feeds. New York: Marcel Dekker, Part 2, pp 319–340
8. Bray WJ (1977) The processing of leaf protein to obtain food-grade products. In: Wilkins RJ (ed), Green Crop Fractionation (Proc occasional simp no 9). British Grassland Society, pp 107–117
9. Bray WJ, Devadas RP, Kamalanathan G, Vijayalakshmi P (1976) Leaf nutrient feeding program in Coimbatore, India. Proc 1st Int Congress on Eng and Food, Boston, pp 17–25
10. Bray WJ, Humphries C (1978) A comprehensive bibliography with special reference to the potential use of leaf protein concentrates in human food. Dept Food Science, University of Reading, Reading
11. Bray WJ, Humphries C (1978) Solvent fractionation of leaf juice to prepare green and white protein products. J Sci Fd Agric 29:839–846
12. Bray WJ, Humphries C, Ineritei MS (1978) The use of solvents to decolorise leaf protein concentrates. J Sci Fd Agric, 29:165–171
13. Bray WJ, Humphries C (1979) Preparation of white leaf protein concentrate using a polyanionic flocculants. J Sci Fd Agric 30:171–177
14. Buchanan RA (1969) In vivo and in vitro methods of measuring nutritive value of leaf protein preparation. Br J Nutr 23:533–545

15. Bunchanan RA (1969) Effect of storage and lipid extraction on the properties of leaf protein. J Sci Fd Agric 20:359–364
16. Cyanamid (1976) Superfloc flocculants for water treatment. Bulletin IND 146-2 Rev 4/76, American Cyanamid Company, NJ
17. De Fremery D, Miller RE, Knuckles BE, Bickoff EM, Kohler GO (1974) Soluble protein from alfalfa. Proc 12th Techn Alfalfa Conference, Berkeley, ARS-USDA, pp 104–109
18. De Fremery D, Kohler GO (1978) Progress towards production of soluble food-grade LPC. 2nd Int Green Crop Drying Congress, Saskatoon, Saskatchewan, 20–25 August
19. Doraiswamy TR, Singh N, Daniel VA (1969) Effects of supplementing ragi diets with lysine or leaf protein on the growth of children in India. Br J Nutr 23:737–743
20. Edwards RM, Miller RE, De Fremery D, Knuckles BE, Bickoff EM, Kohler GO (1975) Pilot plant production of an edible white fraction leaf protein concentrate from alfalfa. J Agr Food Chem 23:620–626
21. Ellis RJ (1977) Chloroplast proteins and their synthesis. In: Norton G (ed), Plant Proteins. London: Butterworths, pp 25–40
22. Fiorentini R, Pachetti G, Galoppini C (1978) Ricerche e realizzazioni europee nel settore del frazionamento umido delle erbe. Atti Acc Ec Agr Georgofili 25:93–135–161
23. Fiorentini R, Lepidi AA (1980) La situazione italiana nel settore del frazionamento umido dei foraggi. Agric Ital 109:315–328
24. Fiorentini R, Galoppini C (1981) Pilot plant production of an edible alfalfa protein concentrate. J Food Sci 46:1514–1517
25. Galoppini C, Fiorentini R, Pachetti G (1978) Leaf protein concentrates for animal feeding: Italian research and realization. III Congreso Mundial de Alimentacion Animal, Madrid, 23–27 October.
26. Gerloff ED, Lima IH, Stahmann MA (1965) Aminoacid composition of leaf protein concentrates. J Agr Food Chem 13:139–143
27. Hove EL (1975) Green leaves as a source of food. New Zealand J Dairy Sci Tech 10:138–141
28. Hove EL, Bailey RW (1975) Towards a leaf protein concentrate industry in New Zealand. New Zealand J Exp Agric 3:193–198
29. Ineritei MS (1976) The effects of solvents on the properties of leaf protein concentrates. M Sc Thesis, University of Reading
30. Knuckles BE, Edwards RH, Kohler GO, Whitney LF (1980) Flocculants in the separation of green and soluble white protein fractions from alfalfa. J Agr Food Chem 28:32–36
31. Koegel RG, Bruhn HD, Kai Kong Ng, Ream HW (1976) Plant juice protein and moisture expression from organic materials. A bibliography. Res Div College of Agric and Life Sciences, University of Wisconsin, Madison
32. Kohler GO, Knuckles BE (1977) Edible proteins from leaves. Food Techn 31: 191–195
33. Kung SD, Saunder JA, Tso TC, Vaughan DA, Womak M, Staples RC, Beecher GR (1980) Tobacco as a potential food source and smoke material: nutritional evaluation of tobacco leaf protein. J Food Sci 45:320–325
34. Lepidi AA, Fiorentini R (1981) Cibo ed energia dall'agricoltura: un conflitto risolvibile. Giornale di Agricoltura 91 (8):40–41
35. Lepidi AA, Fiorentini R, Nuti MP, Galoppini C (1981) Protein and energy from wet green crop fractionation. Int Conference on Energy Use Management, Berlin, 26–30 October
36. Lyttleton JW (1973) Proteins and nucleic acids. In: Butler GW, Bailey RW (eds), Chemistry and Biochemistry of Herbage. London: Academic Press, pp 63–105
37. Miller RE, De Fremery D, Bickoff EM, Kohler GO (1975) Soluble protein concentrate from alfalfa by low temperature acid precipitation. J Agr Food Chem 23: 1177–1180
38. Olatunbosun DA, Adadevoh BK, Oke OL (1972) Leaf protein: a new protein source for the management of protein calorie malnutrition in Nigeria. Nigerian Med J 2:195–199

39. Ostrowski-Meissner HT (1980) Quantities and quality of protein extracted from pasture herbage using heat precipitation or ultrafiltration procedures. J Sci Fd Agric 31:177–187
40. Pirie NW (1971) A survey of other experiments on protein production. In: Pirie NW (ed), Leaf Protein; its Agronomy, Preparation, Quality and Use. Oxford: Blackwell Scientific Publications, pp 29–43
41. Pirie NW (1978) Leaf protein and other aspects of fodder fractionation. London: Cambridge University Press
42. Pirie NW, Telek L (1980) Leaf protein newsletter. Rothamsted Experimental Station, England
43. Saunders RM, Connor MA, Booth AN, Bickoff EM, Kohler GO (1973) Measurement of digestibility of alfalfa protein concentrates by in vivo and in vitro methods. J Nutrit 103:530–535
44. Shah FH (1971) The effect of solvents on the extractability of lipids from leaf proteins. Pakistan J Sci Ind Research 14:207–215
45. Singh N (1971) Feeding trials with children. In: Pirie NW (ed), Leaf Protein: its Agronomy, Preparation, Quality and Use. Oxford: Blackwell Scientific Publications, pp 131–134
46. Subba Rau MS, Singh N, Prasannapa G (1967) Preservation of wet leaf protein concentrates. J Sci Fd Agric 18:295–298
47. Tao M, Boulet M, Brisson GJ, Huang KH, Riel RR, Julien JP (1972) A study of the chemical composition and nutritive value of leaf protein concentrates. Canadian Inst Food Sci and Techn J 5:50–54
48. Tragardh C (1978) Leaf protein concentrate. Processing including ultrafiltration and isopropanol extraction, and its influence on product properties. 5th Int Cong Food Sci Techn, Kyoto, 5–8 September
49. Wang JC, Kinsella JE (1976) Functional properties of novel proteins: alfalfa leaf protein. J Food Sci 41:286–292
50. Wang JC, Kinsella JE (1976) Functional properties of alfalfa leaf protein: foaming. J Food Sci 41:498–501
51. Waterlow JC (1962) The absorption and retention of nitrogen from leaf protein by infants recovering from malnutrition. Br J Nutr 16:531–540
52. Wildman SG (1974) Tobacco and other species of nicotiana plants as a potential source for large scale production of a pure protein for human consumption. Mimeographed Report, University of California, Los Angeles
53. Yen Chen Yen (1976) Acrylamide and polyacrylamide. SRI, Report no 99, Menlo Park, California

Protein extraction from tobacco leaves: technological, nutritional and agronomical aspects

PAOLO FANTOZZI and ALESSANDRO SENSIDONI

Istituto di Tecnologie Alimentari, Università degli Studi di Udine, P. zza M. Kolbe, 4, 33100 Udine, Italy

Key words: tobacco, protein, technology, nutrition, cultivation, curing

Abstract. The goals of the present studies were: (1) to obtain smoking material from tobacco with reduced toxicity; (2) to use the tobacco leaf, as an alternative to other plants, as a source of food protein. On the grounds of technological, nutritional and agronomical results, the following observations can be made: (1) It is feasible to lower the toxicity of the tobacco smoke by extracting protein from the green leaves before curing. (2) The cured deproteinized tobacco can easily be utilized for the production of smoking material without any important alteration of its taste. (3) The composition of the tobacco protein concentrate is very interesting and is superior to other protein concentrates. (4) Its *in vivo* and *in vitro* digestibility is one of the highest among the vegetable protein sources which have been studied. (5) Use of high planting densities and high levels of nitrogen fertilization per hectare results in high yields of total protein from tobacco (1794 Kg/ha).

Introduction

Until recently, tobacco cultivation was performed with only one goal in mind: that of obtaining a commercial product to be used for smoking. However, at the same time, considerable effort has gone into research in attempts to obtain a product that is of better quality and less harmful. For this purpose, it is extremely important to focus on a technique that includes a series of operations that tend to diminish the quantity of those substances present in tobacco that give rise, after combustion or pyrolysis when smoked, to substances that are considered to be cancerogenic or precursors of cancerogenic compounds (nitrous amines, benzopyrenes, ^{210}Po, phenols, proteins, etc.) [27, 30, 31, 34]. For example, the separation by pressing of the juice from leaves that is liberated in the course of homogenization of the leaves, brings about a drastic and favorable decrease in protein, alkaloids, and polyphenols in the fibrous residue.

In some of our previous reports [8, 10, 11, 12], some general results were presented which were related to the possibility of using the by-products which were obtained by the American HCL method used for preventive deproteinization in the production of smoking tobacco [4, 31, 32]. This was done with the hope of the possible recovery of protein materials of high nutritional value (Figure 1). More recently, however, we have analyzed the

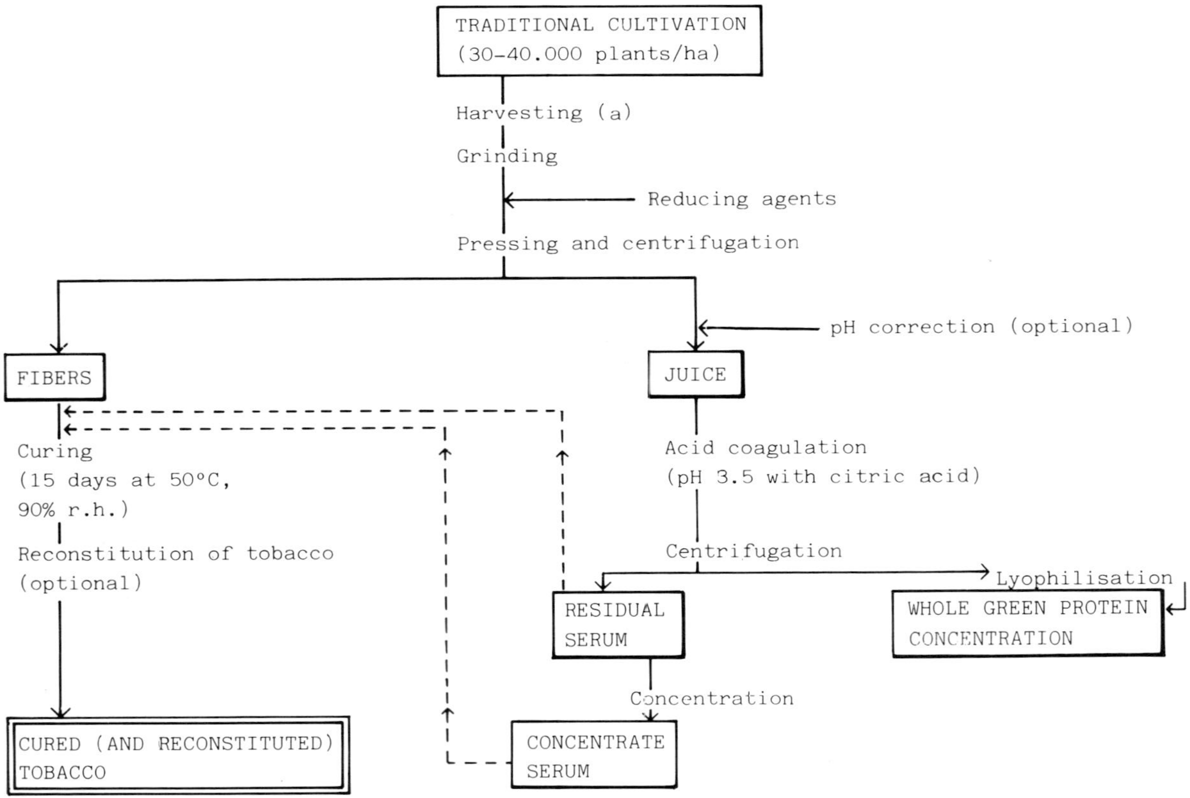

Figure 1. Technological protocol for obtaining deproteinized smoking tobacco.
(a) = 5 days before bud stage.
▭ (double box) = Product; - - → = Facultative operation
▭ = By-product ——→ = Obligatory operation

Table 1. Theoretical (intensive) and real (traditional) availability of protein analyses

	Traditional cultivation (30 000 plants/ha)		Intensive cultivation (400 000 plants/ha)	
	Italy	EEC	Italy	EEC
Hectares (× 1000)	64	90	64	90
Dry matter (tons × 1000)	108	181	1344	1890
Protein (tons × 1000)				
(a) total	13	22	256	359
(b) extractable	2.2	3.7	44	61

Table 2. The amino acid composition of ribulose 1–5 diphosphate carboxylase obtained from Virginia Bright tobacco (cospaia BC 60 variety) in comparison to other proteins

	Tobacco	Soy [24]	Egg [17]	FAO/WHO Reference pattern [13]
Isoleucine[a]	4.1	4.9	5.8	4.0
Leucine[a]	8.4	7.7	9.0	7.0
Lysine[a]	5.7	6.1	6.7	5.5
Phenylalanine[a]	6.1	5.4	5.3	–
Tyrosine[a]	3.3	3.7	4.3	–
Phenylalanine + Tyrosine	9.4	9.1	9.6	6.0
Methionine[a]	1.8	1.1	3.0	–
Cysteine	1.6	1.2	2.1	–
Methionine + Cysteine	3.4	2.3	5.1	3.5
Threonine[a]	3.7	3.7	5.3	4.0
Tryptophan[a]	1.1	1.4	1.8	1.0
Valine[a]	5.8	4.8	7.2	5.0
Histidine	1.7	2.5	2.6	–
Aspartic acid	7.9	11.9	10.7	–
Serine	3.6	5.5	7.7	–
Glutamic acid	11.0	20.5	3.8	–
Proline	3.0	5.3	7.7	–
Glycine	4.4	4.0	12.3	–
Alanine	4.5	3.9	–	–
Arginine	4.6	7.8	6.4	–

[a] Essential amino acid.

actual possibility of the utilization of tobacco as a primary source of protein [6, 7, 9]. The theoretical productive capacities in Italy and in the EEC are reported in Table 1.

The proteins of major nutritional interest are those contained in the chloroplasts; they can be separated into 2 fractions: insoluble or structural (50%) and soluble or enzymatic (50%). Among the soluble proteins present in the juice, the enzyme ribulose 1–5 diphosphate carboxylase (or 'protein fraction 1' or 'F_1P') is of particular interest, from a dietary viewpoint because of its amino acid content (reported in Table 2).

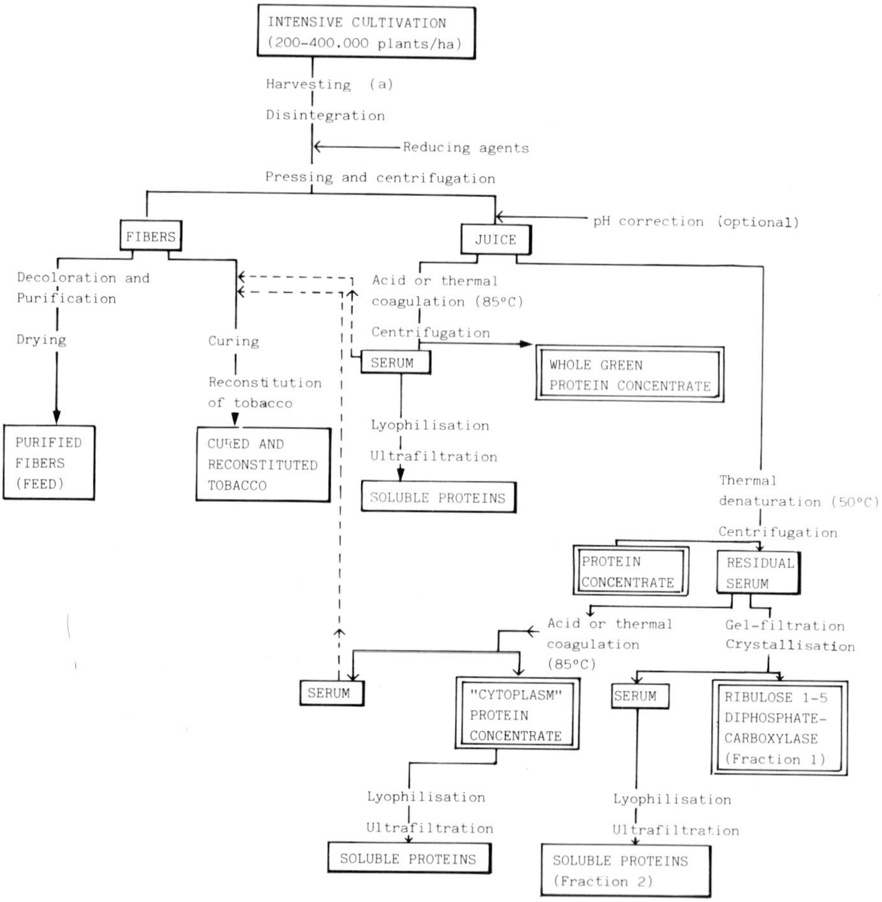

Figure 2. Technological protocol for obtaining protein fractions of different levels of purification from tobacco leaves.
(a) = Time: 5 days before bud stage.
(double box) = Product – –→ = Facultative operation
(single box) = By-product ——→ = Obligatory operation

While there have been limited attempts at direct utilization of tobacco leaves in human and animal nutrition [29], the most widely used technique involves fractionating the green cellular juice. Once the processes of trituration and pressing of the tobacco leaves has been performed, different proteinaceous materials are produced which are in different forms, based on the successive stages of purification:

(a) protein concentrates: chloroplastic and cytoplasmatic (combined or separated).
(b) crystallizable proteins (essentially the above mentioned 'Fraction 1').

A diagram of some possible operations (or protocol) is given in Figure 2.

Table 3. Possible nutritional and dietary uses of the proteins extracted from the tobacco leaf [12, 19]

Nutritional uses	Therapeutic uses
– Additives in non-alcoholic beverages to increase their nutritional value – Milk substitutes for those who cannot metabolize lactose – Integration into some cereal products to improve the amino acid profile	– Nutrition for people with renal insufficiency: the absence of Na^+ and K^+ in these proteins can severely limit the frequency of hemodialysis; possibility of using proteins rather than mixtures of simple amino acids in diets of equal nutritional value, and therefore reducing the volume of the diets – Nutrition for people with gastro-intestinal problems (G-I, esophageal tumors, pylorospasma, etc.): the purity of the protein allows for the use of complete, wholesome liquid diets, with low residue – Nutrition for people who must severely restrict Na^+ and K^+ intake

These products present obviously nutritionally diversified organoleptic characterstics. In particular, they could be utilized as protein integrators in zoological animal feed or in human nutrition as pharmaceutical preparations [6, 35].

These last two aspects are currently of particular interest. The crystallized 'Protein Fraction 1' can be completely soluble in water and is tasteless, odorless, free of potassium and sodium, lipids, glucides, and purine and pyrimidine bases [19, 27]. Some possible uses are listed in Table 3. The serum residue, on the other hand, as was discussed in another report [10] is utilized (after concentration) by combining it with the fibrous residue destined for cigarette production.

In this discussion, we intend to briefly review some results [9, 10, 12] that are of particular practical interest and to consider the optimization of the two different technological processes used for: (1) eliminating and/or reducing undesirable components (proteins, polyphenols, alkaloids) in smoking tobacco, and (2) extracting proteins with the objective of obtaining various protein fractions for possible nutritional use.

Materials and methods

Materials

- The tobacco varieties used were: Erzegovina, Burley and Virginia Bright (Cospaia BC 60). These were traditionally cultivated and furnished by ESAU (Ente di Sviluppo Agricolo nell'Umbria, Perugia, Italy).

Equipment

- Homogenization of samples:
  Universal Aghetto blender (1400–2800 rpm).

- Extraction of tobacco juice:
  Rambotti vertical hydraulic press (max. spec. pressure: 20 atm.).
- Determination of pH and CN: pH meter Orion mod. 50.
- Determination of reducing sugars and of total alkaloids: Technicon Auto-analyzer using the 'Philip Morris' method.
- Determination of static combustion, wet and dry smoke condensate and CO: Filtroma Smoking Machine 300.
- Determination of the humidity of the condensate materials: Karl Fisher apparatus.
- Determination of 'Tar predictor': John Payne Instrument.
- Rapid determination of nitrogen: Kjheltec Tecator PBI Apparatus.
- Nicotine determination: Vapor Current Analyzer.
- Otello Nepi vapor generator.
- Gibertini analytical scale.
- Thermostabile PN3 Orlando Valentini thermostatic incubator.
- Biorad apparatus for disc gel electrophoresis.
- Agrotecno hammer mill (operative capacity 300 Kg/hour).
- Veronesi Separatori vertical centrifuge (with internal disc automatic sediment release).
- Minifast 3400 vacuum freeze dryer, Edwards High Vacuum.
- Luwa industrial concentrator.
- Spectronic 20 colorimeter, Baush and Lomb.

Methods

Storage. The tobacco was stored in a refrigerated (+ 2 °C) or frozen (− 20 °C) state, depending on the particular case, before being subjected to processing or analysis.

Extraction. 400 Kg of freshly-picked tobacco leaves were immersed in a solution of water containing 2 Kg of SO_2 to avoid possible oxidation of the product in the course of subsequent operations. The product was then subjected to triturition with a hammer mill and squeezed in a press to obtain a yield in juice of about 56% and a fibrous residue (Fiber 1). The green juice was immediately centrifuged to separate the fibrous residue (Fiber 2), subsequently coagulated by adjusting the pH to 3.5 with citric acid, and then left to stand for 12 hours at + 5 °C. The protein precipitate was then separated by continuous centrifugation, washed and freeze-dried.

In this preliminary phase of research, separation of the various protein fractions present in the juice was not done. The freeze-dried residue obtained by pressure or centrifugation (Fibers 1 and 2) and the acid serum, concentrated until it was one-sixth of its original volume in an industrial concentrator (descending film process), were used for the production of cigarettes. The procedures followed are part of the technological protocols reported in Figures 1 and 2.

Table 4. Agronomic variables used for evaluation of the protein concentration in the leaves[a]

Varieties:	Cospaia BC 60 Burley Erzegovina
Nitrogen fertilization (kg/ha):	0 92 184 276
Density of planting (plants/ha):	31 250 100 000 166 600

[a]Each variety was tested, at the same time (1980), with 3 densities of planting at each of the 3 levels of nitrogen fertilization.

Table 5. Calender of transplants and harvests (1980 year)

Variety	Date of transplant	Time lapse (in days) from transplant		
		1st harvest	2nd harvest	3rd harvest
Erzegovina	6/6/80	53	99	151
Cospaia	6/6/80	83	148	–
Burley	2/6/80	65	124	–

Fermentation and Curing. The various fractions of tobacco (whole homogenized, fiber and serum) were submitted to fermentation in a controlled environment at 50 °C with a relative humidity of 30% until the product reached a moisture level of 40%. Immediately afterwards, it was cured for 15 days at 50 °C at a relative humidity of 90% in a ventilated incubator (Figure 1).

Cigarette production. Cigarettes were produced from each sample of cured tobacco and then submitted to a jury of three judges from the State Monopoly. The cigarettes were made with a Baby Cigarette Maker after prior sifting of the cured product to separate out the particles that were smaller than 1 mm and larger than 0.5 mm.

Agronomic testing methodology. The agronomic variables used are summarized in Table 4. The preparation of the land and detailed statistical data related to the agronomic tests conducted have been described [9]. Each variable was repeated three times and the land parcels were distributed according to a factorial scheme. A 'border effect' was avoided by excluding from the calculations data for the plants located around the perimeter of each parcel.

Harvesting was performed according to the outline given in Table 5. The

Table 6. Effect of the intensity of triturition of the tobacco on some compositional and organoleptic characteristics of the cured product[a]

	Sample 1 (low triturition)*			Sample 2 (high triturition)**		
	pH	Sugars (g%)	Smoke taste test	pH	Sugars (g%)	Smoke taste test
Homogenized	5.04	10.9	C	4.95	10.4	C
Fiber 1	5.10	–	A	4.95	–	A
Fiber + Serum (1:2) (thermal coagulation)	5.00	10.9	A	4.94	8.7	A
Fiber + Serum (1:2) (acid coagulation)	3.52	9.3	B	3.69	9.0	B

[a]The general method of extraction is reported in Figure 2; * = (30″, 1400 rpm + 30″, 2800 rpm); ** = (60″, 1400 rpm + 60″, 2800 rpm); A = mediocre taste; rather pungent aroma; bitter aftertaste; a papery aftertaste which is typical of the cured midrib of the leaf; rather good combustibility; B = mediocre taste; slightly bitter aftertaste; slightly pungent aroma; papery aftertaste; good combustibility; C = fairly good taste; neutral aftertaste; weak aroma; good combustibility.

Table 7. Effects of protein extraction on tobacco smoke characteristics

Bright tobacco	Nicotine	Predicted smoke condensate (mg/g D.M.)	Smoke taste test
Control (traditional curing)	0.40	45.3	Taste: light, agreeable Combustion: good
Homogenized	0.46	44.3	Taste: neutral, agreeable Combustion: good
Homogenized + deproteinized + concentrated serum	0.38	43.0	Taste: neutral, agreeable Combustion: good
Homogenized and deproteinized[a]	0.16	–	Taste: pungent Combustion: very poor

[a]This product could be utilized as inert support in the commercial blends to lower the final smoke concentration of nicotine.

plants were cut with a sickle at a level 5–10 cm above the ground when the majority of the plants had formed a floral bud [3, 27]. After the first harvest, the land parcels were irrigated and hoed and, on the previously cut stalks, ratoons were grown to obtain a second, and when possible, also a third harvest. After each harvest, a portion of the plants from each parcel was used for different analytic evaluations.

Analytic determinations:

- Dry substance, pH, reducing sugars, total alkaloids, and CN were analyzed according to official methods of analysis [1].
- Nicotine, dry and wet smoke condensate, and CO were analyzed according to the ISO 3400 and 3308 methods [16].
- Static combustion and 'tar predictor' were done according to the current methods of the State Monopoly.
- Total nitrogen was determined by a microchemical method [21].
- Crude protein was calculated by multiplying the value for total nitrogen by the factor 6.25.
- Ribulose 1–5 diphosphate carboxylase levels were determined by the electrophoretic method of Masseroni [22].
- Total polyphenols (performed only on the liquid fraction) were measured colorimetrically with the Folin Ciocalteu reagent expressing the results as gallic acid (mg/g dried substance) [28].
- Polyphenoloxidase (PFO) was measured on the brown juice according to the Montedoro method [23].
- Evaluation of the protein quality was done: (a) with the RPV multi-dose method (Relative Protein Value) [2, 25], and (b) through enzymatic digestion in vitro (EUD; Enzymatic Ultrafiltrate Digest index) [15].

Results with traditional cultivation

Smoking material

The results of major interest with regard to tests performed on the smoking materials are reported in Tables 6, 7 and 8. Table 6 shows the effect of the intensity of triturition of the green tobacco leaf on the final organoleptic characteristics of the cured smoking material. The introduction of this operative variable (intensity of triturition) was justified by the need to extract from the foliate tissue the greatest possible quantity of juice. This was done to obtain the greatest quantity of protein with respect to the structural and organoleptic quality of the residue destined for smoking. Between the samples obtained with low or high degrees of homogenization, no differences which are worthy of discussion are apparent.

However, difficulties arise with the product obtained with elevated triturition when, after the curing process, it is sent to be made into cigarettes. Because the size of the dried particles (comparable to tobacco dust), the filling of the cigarettes becomes particularly difficult. There is an increase in the risk of the partial or even total emptying of the cigarette. To avoid this problem, it becomes necessary to use a process of reconstruction of the triturated tobacco leaves in the technological process. Conversely, when the tobacco is triturated at a low number of revolutions, fibrous particles of sufficiently large size are obtained so as to assure the filling and

Table 8. Modification and evaluation of some analytical parameters during the smoking process

Parameters	Control	Homogenized and deproteinized tobacco + concentrated serum	Effect of protein subtraction
Cigarette weight (g)	1.15	1.27	–
Cured tobacco, humidity (%)	12.00	13.00	–
Static combustion (mg/sec)	0.73	1.18	–
Tobacco: Reducing sugars (%)	14.59	9.00	–
Tobacco: Total alkaloids (%)	0.73	0.5	Positive
Puff number	14.50	10.00	–
Wet smoke condensate (mg/cig)	22.90	13.30	Positive
Dry smoke condensate (mg/cig)	21.20	12.00	Positive
Nicotine in the smoke (mg/cig)	0.66	0.37	Positive
CO in the smoke (mg/cig)	14.00	16.00	Negative
Tar predictor (mg/cig)	39.40	39.60	Null
CN (mg/cig)	150.00	200.00	Negative

Table 9. Percent distribution of proteins in various fractions obtained during the extraction

Leaf (total)	100		
Fiber		68	
Juice		32	
Residual serum			15
Whole leaf protein concentrate			17

holding of the cigarette without difficulties. The effects of varying a few analytic parameters, in the course of the curing of the fibrous residue subjected to different treatments, are reported in Table 7. The technique of homogenization does not seem to greatly influence the qualitative characteristics of the smoking material obtained. However, the removal of leaf juice brings about, at the same time, a marked decrease in the nicotine content, a drastic worsening of the taste and a decrease in the combustibility of the cured product.

Of great interest is the verification that the readdition of the deproteinized leaf juice (serum) to the fibrous residue reestablishes the organoleptic characteristics of the cured product almost totally. This effect is

Table 10. Composition of leaf protein concentrates

Composition		Whole leaf protein concentrate		
		Tobacco	Alfalfa (Polyprotein) [14]	Alfalfa (Proxan) [18]
Moisture	(%)	4	10	–
Ash	(%)	7	12	9
Total N_2	(%)	8	–	–
Protein	(%)	50	50	57
Fiber	(%)	2	2	3
Fat	(%)	8	9	12
Chlorophyll	(mg%)	117.2	–	–
Carotenoids	(mg%)	111.3	150	288
Nicotine[a]	(mg%)	370.0	–	–

[a]Present because of an insufficient purification of the product at this preliminary stage of the project.

notably elevated if the addition is made when the serum is concentrated. These observations suggest the possibility of recycling the serums, which if not used are by themselves quite polluting when released into the external environment after the protein fractions, that represent the highest commercial and nutritional value, have been extracted. The possibility of concentrating the volume before adding it to the fibrous residue allows, in part, a decrease in the energy cost of the curing operations.

Based on the above observations, some investigations were performed comparing Virginia Bright tobacco that was only homogenized with the same tobacco that was deproteinized and added to the serum acid concentrate. The results are reported in Table 8. Deproteinization has a positive effect on the total alkaloid, wet and dry smoke condensate and nicotine content. A cause of the positive decrease of the above parameters is the increase of the velocity of combustion of the cigarette filled with homogenized tobacco with a subsequent decrease in the number of puffs per cigarette. A slightly negative effect of deproteinization, however, was the increased contents of CO and CN in the smoke. These results confirm what has been observed earlier by other authors [30, 31, 32, 34] regarding the positive effect that the removal of proteins from tobacco leaves has on the total toxicity of smoking.

Protein concentrate

As previously mentioned, an attempt was made to extract the proteins from tobacco leaves on a semi-industrial level with the following objectives: (1) furnishing deproteinized material for the production of cigarettes (the results relating to this, were discussed in the preceding section), (2) evaluation of the total yield of the extracted protein material, and (3) evaluation of the biological value of the whole protein concentrate. The operations followed were part of the technological protocols reported in Figures 1 and 2.

Extraction yields. In Table 9, the yields of the industrial tests of the extraction of the tobacco proteins are reported. The structural and enzymatic proteins, present in the protein concentrate obtained by the industrial process used, represent only 17% of the total proteins present in the tobacco leaf. It is appropriate to consider why 68% of total protein remains available in the fiber while only 32% passes into the juice. The low extraction yields obtained are due to the effect of the gentle breaking up of the leaves used in these tests during the triturition phase. This was done to maintain a sufficient structural conformation of the deproteinized fibrous residue to allow it to be used in the production of smoking tobacco [10]. In the case where the tobacco is to be used principally for protein production, it would be necessary to produce a more finely homogenized product of the fresh foliate material.

Evaluation of the whole protein concentrate. The composition of the protein material obtained is reported in Table 10. In general the proximate composition of the protein concentrate of tobacco is very similar to that of alfalfa. A difference exists in the minor concentrations of lipids and carotenoids and in the presence of nicotine (the latter due to insufficient purification of the material used in this preliminary phase).

An evaluation of the biological quality of the protein concentrate is reported in Table 11. A careful analysis of this data shows that the RPV and EUD values are, for the most part, comparable or superior to those of the other plant protein sources examined, demonstrating the high biological values of the tobacco leaf product. In particular, for the EUD data, tobacco surpasses seeds and, for the proteins extracted, it is equal to nettle and inferior only to one of the alfalfa protein concentrates (note that this alfalfa concentrate was submitted to further purification to separate out the chloroplastic fractions). These results are in agreement with those obtained by other authors [19, 29].

Results with intensive cultivation

Varieties

From an examination of the data in Table 12, it is evident that the most productive variety, at the maximum density of planting, and with fertilization of 200 units of N/ha, is the Burley variety, followed by Erzegovina and Cospaia. It is worth considering, however, that the Cospaia variety surpasses the quantity of proteins produced by Erzegovina if one considers only two harvests. Some agronomic results, obtained in a similar study conducted in 1979, confirmed this pattern [26].

Nitrogen fertilization

The effects of varying the level of nitrogen fertilization on the yields of protein per hectare for the Erzegovina variety in the course of 3 successive

Table 11. Protein quality evaluation of tobacco protein concentrate in comparison to other proteinaceous products

Protein source	Protein quality in vivo (RPV) [2]	Protein quality in vitro (EUD) [14]
Leaves		
Tobacco (Whole green protein concentrate)	63.4	61.0
Alfalfa (whole green protein concentrate)	48.6	–
('cytoplasm' protein concentrate)	–	79.3
Nettle (whole green protein concentrate)	–	63.6
Seeds		
Sunflower	52.1	54.7
Pea	63.0	–
Chick Pea	100.0	–
Broad Bean	–	58.5
Lupin	–	51.9
Grapeseed	–	39.2
Standard proteins	100[a]	100[b]

[a] Lactalbumin.
[b] Egg.

Table 12. Agronomical results: influence of variety[a]

Variety	Harvest	Burley (Kg/ha)	Burley (RSD)[a]	Erzegovina (Kg/ha)	Erzegovina (RSD)	Cospaia (Kg/ha)	Cospaia (RSD)
Dry matter	1st	4681	8.98	3016	14.50	5147	2.45
	2nd	4017	2.37	2284	6.93	3206	12.79
	3rd	–		1423	3.27	–	
		8698		6723		8353	
Total proteins[b]	1st	840		673		872	
	2nd	841		539		551	
	3rd	–		307		–	
		1681		1519		1423	
F_1P[b]	1st	32		86		95	
	2nd	110		31		30	
	3rd	–		11		–	
		142		128		125	

[a] N = 200 Kg/ha; density = 166.600 pla/ha; RSD = Relative Standard Deviation.
[b] Data are averages of six replicate determinations.

Table 13. Agronomical results: influence of nitrogen fertilization [a]

Nitrogen (Kg/ha)		0		100		200		300	
	Harvest	(Kg/ha)	(RSD)[a]	(Kg/ha)	(RSD)	(Kg/ha)	(RSD)	(Kg/ha)	(RSD)
Dry matter	1st	2734	4.24	2885	10.02	3016	14.50	3213	0.56
	2nd	1719	11.02	1991	0.97	2284	6.93	2929	0.69
	3rd	1477	5.80	1254	2.06	1423	3.27	1530	6.29
		5938		6130		6723		7672	
Total proteins[b]	1st	395		575		673		795	
	2nd	311		402		539		690	
	3rd	284		235		307		309	
		990		1212		1519		1794	
F_1P[b]	1st	48		102		86		–	
	2nd	17		19		31		33	
	3rd	22		13		11		14	
		87		134		128		?	

[a] Variety = Erzegovina; density = 166.600 pla/ha; RSD = Relative Standard Deviation.
[b] Data are averages of six replicate determinations.

harvests are reported in Table 13. A positive effect of nitrogen fertilization on the production of crude protein was observed. This agricultural practice, traditionally avoided for obtaining tobacco to be used for smoking, permits, for doses of 300 units of N/hectare, the doubling of protein produced per surface unit. Identical results were observed in other varieties tested.

Planting density

Table 14 shows the positive effects of increasing the number of plants per hectare on the production of protein for the Erzegovina variety. In some research [36], a planting density of 400 000 plants/hectare has been used. An example is reported in Table 15.

Some industrial tests for extraction were conducted by us [26] and the results achieved either experimentally or through extrapolation, are reported in Table 16. Under our conditions, it was impossible to obtain the same quantity of protein per hectare as that obtained by Wildman [36]. The lower protein concentration in the juice, and therefore the lower extraction level of protein that we found, is due to the limited homogenization of the leaf that we used in our technique, which is described in another note [10]. However, it is useful to recall, at this point, that very high density cultivation trials (between 1.8 and 3.7 million plants/hectare [20]) have always produced yields of protein between 1500 and 2500 Kg/hectare, which is comparable to our results using only 166 600 plants/hectare. Our Institute is presently conducting agronomic trials to definitively verify the possibility of producing quantities of proteins similar to those postulated

Table 14. Agronomical results: influence of transplantation density [a]

Plants (Number/ha)		31 250		100 000		166 600	
	Harvest	(Kg/ha)[a]	(RSD)	(Kg/ha)	(RSD)	(Kg/ha)	(RSD)
Dry matter	1st	3232	13.72	3984	9.65	5147	2.45
	2nd	1586	3.89	2264	2.84	3206	12.79
		4818		6248		8353	
Total protein[b]	1st	533		680		872	
	2nd	297		328		551	
		830		1008		1423	
F_1P[b]	1st	61		51		95	
	2nd	10		6		30	
		71		57		125	

[a]Variety Cospaia, N = 200 Kg/ha; RSD = Relative Standard Deviation.
[b]Data are averages of six replicate determinations.

Table 15. Theoretical yields of protein from extraction of Burley tobacco obtained by Wildman et al. [36][a]

	Dry weight (Kg/ha)	Protein (Kg/ha)
Plants		
Minimum yield	15 000	3600
Maximum yield	22 200	5900
Extraction yields		
Chloroplastic protein concentrate	2250	1333
F_1P	320	320
F_2P	1300	1300
Serum	4049	0

[a]Data obtained by calculation hypothesizing 2 seasonal harvests with two sprouts (cuts) from the parcel tested, utilizing the real production of only one harvest

by Wildman [36]. It should also be mentioned that some authors [5] have verified experimentally that leaf protein content, when calculated from total nitrogen values, is often overestimated. It is worthwhile noting the technological difficulties connected with the seeding or the transplanting of such a high number of plants into an open field. Other experiments will have to be conducted to deal with this problem and to obtain adequate solutions.

Production of Ribulose 1–5 Diphosphate Carboxylase

In Tables 12, 13 and 14 the effects of planting density, nitrogen fertilization and the variety of tobacco on the cellular production of the enzyme ribulose

Table 16. Protein yields from Burley tobacco: experimental (A) and theoretical (B) values

	(A) (density 166 600)[a,b]		(B) (density 400 000)[c]	
	Dried matter (Kg/ha)	Protein (Kg/ha)	Dried matter (Kg/ha)	Protein (Kg/ha)
Leaf	8698	1681	20 961	4051
Fiber	7144	1143	17 216	2755
Juice	1554	538	3745	1296
of which:				
Chloroplastic concentrate	504	252	1214	607
F_1P	51	34	124	82
Serum (including the F_2P)	999	252	2407	607

[a] Based on trials with three consecutive harvests for each parcel.
[b] Nitrogen fertilization level = 200 Kg/ha.
[c] Linear extrapolation from the data from column (A) (density 166 600 plants/ha); for description of the technological protocol, see Figure 1.

1–5 diphosphate carboxylase (F_1P) are also reported. These results overlap perfectly with those of total nitrogen (crude protein) reported previously. Moreover one can observe how the concentration of the F_1P fraction represents, on the average, 7–10% of the total crude protein.

Conclusions

From the large volume of data at our disposition, a few fundamental points of great interest become evident for the future of this research and for its subsequent application in industry:

(1) The agricultural manager in the tobacco sector now finds himself with two possibilities for utilizing his crop: for smoking and as a source of protein. This fact is of extreme importance for a crop that is only recently recognized as heading into a crisis (high cost of labour, decreased demand for cigarettes, etc.). The possibility of having alternative uses for tobacco can be a major advantage for those who, traditionally cultivating this crop, have had poor financial returns. It is worthwhile to realize that, while a lot is known regarding the technology and economics of the first use (smoking tobacco), less is known about the economics concerning the second possible use (the utilization of tobacco as a source of proteins).

(2) It is necessary that all the doubts, associated with the concept of transplanting or seeding a high number of plants per hectare (300 000–400 000), which appears to be necessary for obtaining a sufficient quantity of protein, are evaluated and resolved. Moreover, it is also desirable that studies be done utilizing a wider geographical area in order to find the areas best suited for this particular type of cultivation; areas that provide, among other things, numerous annual harvests.

(3) The problems connected to the technology of protein extraction from tobacco (apart from those already discussed in other studies [12]) are almost solved with respect to obtaining concentrated proteins for use as animal feed from the wastes of the serum and of the residues of the technique. While it has already been studied in the laboratory, the problem of the industrial production of proteins at different levels of purification for human nutritional use, in view of the complete utilization of tobacco in this area, remains to be solved. On this subject, another topic that must be dealt with is that of foreseeing the utilization of the fibrous residue and the deproteinized serum for cattle feed (with prior purification) or for the production of biogas (methane) or ethanol. Studies in that area are presently being conducted in laboratories in and outside of Europe [27, 29].

Acknowledgements. The collaboration of the following Agencies has aided in the realization of this study:
- CNR National Research Council, Rome, Italy: Application-directed projects: 'Research on New Protein Sources and Novel Food Formulations'. 'Agricultural Mechanization'.
- ESAU The Agency for Agricultural Development in Umbria, Perugia, Italy.
- MS State Monopoly, Chemical and Technological Research Service, Rome, Italy.

References

1. A.O.A.C. (1968) Official Methods of Analysis
2. Corco Benedetti P, Tagliamonte B, Bises G, Gentili V, Spadoni NA (1981) Evaluation of protein quality. Am J of Nutrition and Metabolism 25:48–58
3. DeJong DW, Lam JJ (1979) Application of homogenized leaf curing to protein recovery and to the alteration of leaf chemistry for production of less hazardous tobacco. Tob Res 5(1):1–27
4. DeJong DW, Lam S, Lowe R, Yoder E, TSO TC (1975) Homogenized leaf curing 2. Bright tobacco. Beitr Zur Tabak 8:93–101
5. Dumery B, Albo JP, Chouteau J (1981) Approche analytique des proteines de tabac. Implications technologiques. Under Press, SEITA (Institut Expérimental du Tabac) Bergerac, France
6. Fantozzi P (1980) Quando il tabacco non va solo in fumo. Giornale dell'Agricoltura, 90(22):48–49
7. Fantozzi P (1980) La via italiana del tabacco. Salve 4(2):65–67
8. Fantozzi P (1980) Considerazioni introduttive sul recupero di proteine da foglie di tabacco e sulla loro possibile utilizzazione alimentare. Tecn Alim 1:31–35
9. Fantozzi P, Garofolo L, Sensidoni A (1982) Utilizzazione del tabacco come fonte alimentare: Nota III. Influenza di alcune variabili agronomiche sul contenuto in proteine strutturali ed enzimatiche. Industrie Alimentari 195(6):482–488
10. Fantozzi P, Neri M, Sensidoni A (1982) Utilizzazione del tabacco come materiale da fumo e come possibile fonte alimentare: Nota I. Valutazione del tabacco da fumo deproteinizzato. Industrie Alimentari 193(4):277–282
11. Fantozzi P, Sensidoni A (1981) Utilizzazione del tabacco come materiale da fumo e come possibile fonte alimentare. Proceedings of Congress: 'Problemi e prospettive della tabacchicoltura toscana' Arezzo, marzo 1981
12. Fantozzi P, Sensidoni A (1982) Utilizzazione del tabacco come materiale da fumo e come possibile fonte alimentare: Nota II. Valutazione della frazione proteica. Industrie Alimentari 194(5):375–383
13. FAO/WHO (Food and Agriculture Organization/World Health Organization.)

(1973). Energy and Protein Requirements. WHO Tech Rep Series No 522, FAO Nutrition Meetings Report No 52. FAO/WHO, Geneva
14. Fiorentini R, Pachetti G, Galoppini C (1978) Attualità e prospettive del frazionamento umido dei foraggi. Istituto di Industrie Agrarie, Università di Pisa
15. Floridi A, Fidanza F (1975) Valutazione della qualità proteica mediante digestione enzimatica in vitro. S and TA-NU 5:13
16. I.S.O. (1975) International Organization for Standardisation Methods.
17. Jones A (1974) Proteins in diet. In: Jones A (ed), World Protein Resources. Lancaster and London: MTP Publ, page 26
18. Kohler GO, Edwards RH, Lyon CK, McDonald D (1980) Commercialization of the Pro-xan Process for the Production of Leaf Protein Concentrate. A.C.S. Society North Am Chem Congress, Las Vegas, NV, pp 1–14
19. Leaf Protein International Inc (1982) Raleigh N Carolina. The Story of Proleaf
20. Long RC (1970) Tobacco Production for Protein Projects No 03245, 50–62. North Carolina State University Raleigh NC
21. MAF (1980) Metodi Ufficiali Italiani di analisi (Istituto Poligrafico delle Stato) Rome
22. Masseroni V (1978–79) Isolamento e purificazione della D – Ribulosio 1–5 Difosfato Carbossilasi presente nelle foglie di Lemna minor. Ph D Thesis Università degli Studi di Pavia, Italy
23. Montedoro G (1969) Ricerche sulla polifenolossidasi delle uve: II estrazione, purificazione e caratterizzazione dell'enzima. Industrie agrarie, 7:259–267
24. Pompei C, Lucisano M (1974) Purificazione delle proteine di soia per ultrafiltrazione diafiltrazione. La Riv Ital Sost Grasse, 51:149–154
25. Samons KW, Hegested DM (1977) Animal Biossay. In: Bodwell CE (ed), Evaluation of Proteins for Human. Westport: AVI Press, pp 68–80
26. Sensidoni A, Garofolo L, Fantozzi P (1981) Utilizzazione del tabacco come possibile fonte alimentare. Nota 2: effetto di alcune variabili agronomiche sul contenuto proteico delle foglie. Proceedings of 'Giornata tecnica della tabacchicoltura italiana', CNR-ESAU Foligno 25 maggio, Italy
27. Sheen SH (1982) University of Kentucky, Lexington. Unpublished data
28. Singleton VL, Sullivan AR, Kramer C (1971) An analysis of wine to indicate ageing in wood or treatment with wood chips or tannic acid. Am J of Enol Vitic 22:161–170
29. Staron T (1981) Technologies agroalimentaires INRA-Lucé; 317–342
30. Tso TC (1972) The potential for producing safer cigarette tobacco. Agric Sci Rev 10:1–10
31. Tso TC (1977) Tobacco as Potential Food Source and Smoke Material. Beitr Zur Tabak, 9(2):63–66
32. Tso TC, Gori GB (1972) Effect of tobacco characteristics on cigarette smoke composition. In: Schmeltz I (ed), The Chemistry of Tobacco and Tobacco Smoke. New York: Plenum Publ Corp pp 51–63
33. Tso TC, Lowe RH, DeJong DW (1975) Homogenized leaf curing I. Theoretical basis and some preliminary results. Beitr Zur Tabak 8:44–51
34. Usda-Sea USA, (1977) Technical Bulletin No. 1551
35. Wildman SG (1978) Il tabacco é nutriente. Salve 3(12):58–63
36. Wildman SG, Kwanyuen P, Ershoff BH (1977) Production and Biological Evaluation of crystalline fraction I protein from Tobacco leaves. Proceedings of 'Non Conventional Proteins and Foods Conference', Madison Wisconsin

Protein functionality and its relation to food microstructure

ANNE-MARIE HERMANSSON

SIK, The Swedish Food Institute, Box 5401, S-402 29 Göteborg, Sweden

Key words: protein, functionality, soy protein, blood plasma protein, gluten, microstructure

Abstract. During food preparation, structures are formed which give the food its characteristic properties. The addition of protein ingredients is meant to improve the structure and give a positive contribution to functional properties such as texture, and fat- and waterbinding. A functional characterization provides information on the behaviour of a protein in a food system. Factors to be considered in a functional characterization are discussed, and examples are given from studies of various protein ingredients.

Waterbinding and texture characteristics generated by gel formation and their relation to changes in the microstructure are illustrated by results from studies of plasma protein gels. The results clearly show the difference in the gel characteristics as well as the importance of a proper choice of methodology and definition of the property to be studied.

A comparison between microstructure and functional properties of commercially processed and mildly prepared soy protein isolates demonstrates differences caused by processing and their impact on the application of the protein. Results from a study on gluten show that other components than protein may be responsible for drastic changes in the functional properties of a protein ingredient during processing.

Introduction

Proteins, as well as starch and other polysaccharides, are added to foods in order to enhance functional properties such as texture, fat- and waterbinding, foam, and emulsion stability.

A specific food product can be made by using a variety of recipes and processes, which differ in their demand on the functional ingredient(s). Furthermore, proteins with quite different functional properties may give the same final contribution to the product quality, even if it was achieved in different ways. Thus soy protein isolate, sodium caseinate, and blood plasma may all give a positive contribution to a frankfurter. The soy protein isolate achieves this mainly by means of its swelling and, in some cases, gelling properties. Caseinate possesses emulsifying properties and high viscosity and lacks the ability to form heat-induced gels. Blood plasma has a very low viscosity but forms firm gels on heating.

The complexity of the various ways in which a protein ingredient may contribute to a food structure as well as the importance of the recipe and the process of food production, make a functional characterization of a new protein ingredient difficult. The first point to be considered is:

What functional properties are of importance for the application of a protein in a given system?

This question forms a basis for functional characterization, since functional properties can be defined as those physico-chemical properties giving information on how a protein will behave in a food system [1]. It is time consuming, and may even be impossible, to find this answer from studies of complex foods without any previous knowledge of the protein to be tested [2]. From studies in simpler model systems, the functional properties of a new protein ingredient can be identified, and the results used for appropriate design of large scale tests of complex foods.

Demands on the methods used for functional characterization

Before a functional characterization is made, a critical evaluation of the tools is appropriate, so the methods and experimental conditions chosen really provide useful information for the application of the protein in a given system. Furthermore, the relevance of the methodology and terminology has to be considered. Physical properties given by a certain state of aggregation are often confused with each other, since one property may influence the measurement of another property. In this discussion, the term 'gel strength' will be critically examined with regard to texture and waterbinding characteristics.

Relationships between microstructure and functional properties

A better understanding of various functional properties can be obtained from knowledge of the structure. The purpose of food processes such as the production of bread, cheese, butter, sausages etc., is to create a structure which gives the food item its characteristic properties. Therefore, there is an increasing demand for knowledge about food structures and relationships between microstructure and functional properties.

Effects of processing and composition on structure and functional properties

In addition to the major points of functional characterization stated above such as an identification of relevant properties, the proper choice of methodology and an understanding of how these properties were achieved, some other factors are important to consider in the evaluation of new protein ingredients.

Firstly, commercially produced protein isolates may differ considerably from mildly prepared isolates with regard to functional properties, due to physical or chemical modifications during the preparation process. Therefore, results obtained from studies of native protein may not be valid for commercially available products from the same raw material.

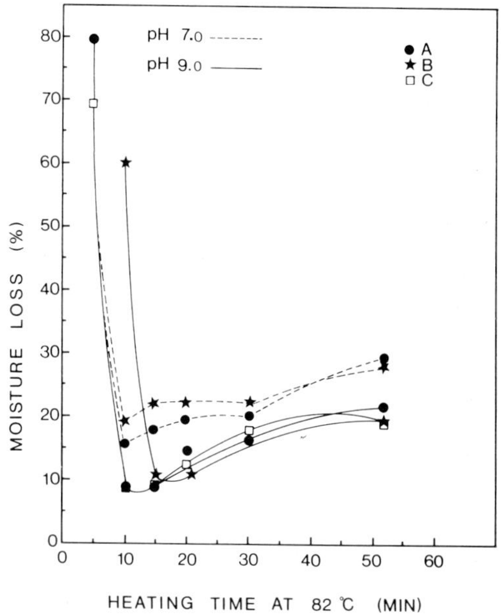

Figure 1. Moisture loss as a function of heating time at 82°C of 5% plasma protein gels at pH 9.0 and 7.0 according to the modified net test. A, B and C refer to different batches of plasma [3].

Secondly, other components than proteins may contribute significantly to the function of a protein ingredient. This point may be especially important to consider in the case of protein concentrates with their lower protein contents than isolates.

Examples will now be given of the various aspects of functional characterization observed in studies of different types of protein gel systems.

Properties generated by gel structures

The ability of proteins to form a continuous network structure is important in most half-solid and solid foods, such as milk, meat, fish, egg, and cereal products. The gel structure contributes to the characteristic texture of a product and acts as a matrix, holding water, as well as fat and other components.

The term 'gel strength' is often used to describe gels, without closer definition. Sometimes it refers to waterbinding and sometimes to special texture characteristics. It may even include both waterbinding and texture without distinction.

A systematic study was made of gels prepared under a wide range of conditions [3,4, 5]. Waterbinding and texture characteristics were measured and methodology aspects considered. In the first part of the study, blood plasma was used because of its relative simplicity of handling but the gel properties discussed and their relation to microstructure are of a general nature.

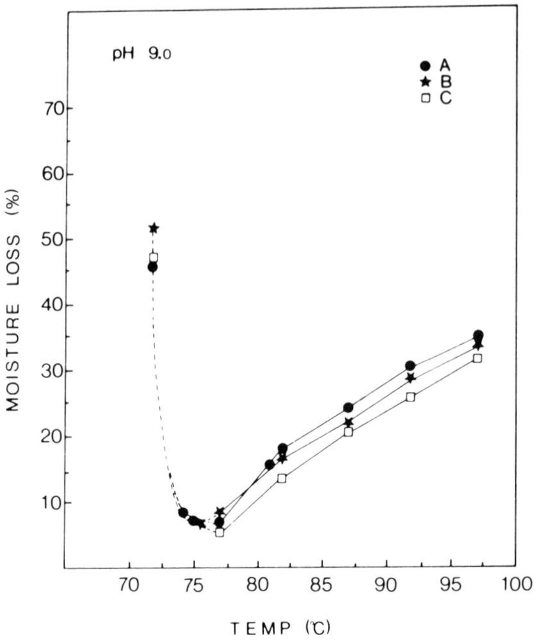

Figure 2. Moisture loss as a function of heating temperature of 5% plasma protein gels at pH 9.0 according to the modified net test. A, B, and C refer to different batches of plasma [3].

Vegetable protein products are often more difficult to evaluate, and special problems regarding soy protein isolates and gluten will be discussed later.

Figure 1 shows moisture loss as a function of heating time at 82°C of 5% plasma protein gels at pH 9.0 and 7.0. Frozen and thawed plasma has a pH value close to pH 9.0 before adjustment. It is seen that the moisture loss (ML) passes through a minimum after 10–15 min of heating.

Figure 2 and 3 show moisture loss of 5% plasma protein gels at pH 9 and 7 as a function of heating temperature. At temperatures below 73°C the gels were too poor for reliable results to be obtained. The moisture loss (ML) passed through a minimum, which can be seen in Figure 2, and increased markedly at temperatures above 80°C at both pH values.

The results shown indicate that the ability of the gel to hold water had an optimum close to the gelation temperature. Further heat treatment resulted in an increased moisture loss. An observation of great importance for the use of blood plasma as a functional ingredient is that a heating temperature of at least 75°C has to be used in order to take advantage of its gelling properties.

Texture is the other important property generated by gel formation. It is a complex concept and numerous measures can be used to describe various texture characteristics. Only a few examples can be given in this paper.

Figure 4 and 5 show force compression curves of 5% plasma protein gels at

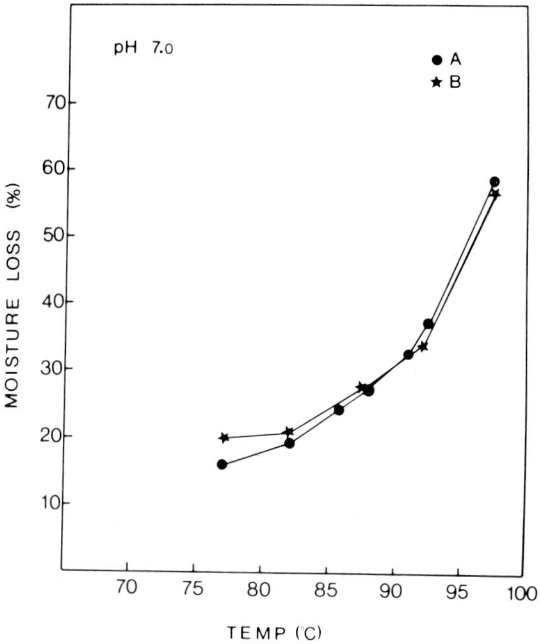

Figure 3. Moisture loss as a function of heating temperature of 5% plasma protein gels at pH 7.0 according to the modifed net test. A and B refer to different batches of plasma [3].

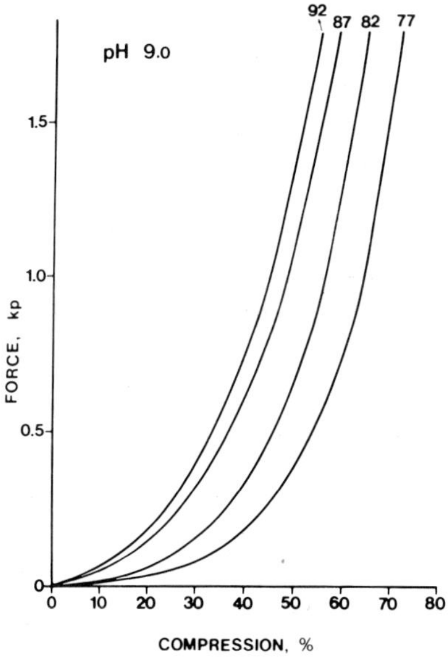

Figure 4. Force compression curves of 5% plasma protein gels at pH 9.0 as a function of heating temperature [4].

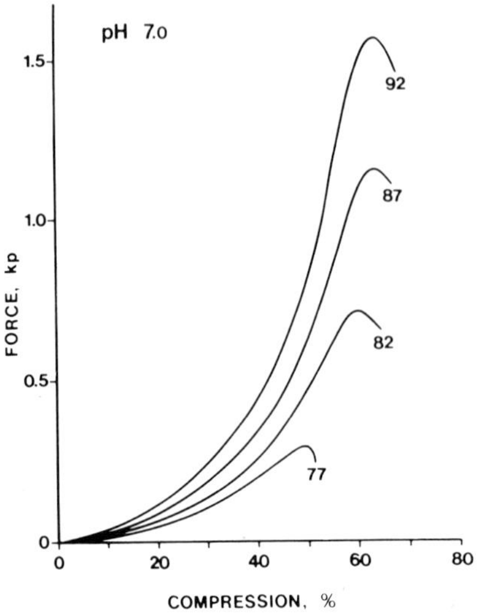

Figure 5. Force compression curves of 5% plasma protein gels at pH 7.0 as a function of heating temperature [4].

pH 9.0 and 7.0 as a function of heating temperature. The compression force increased with heating time at all degrees of compression both at pH 9 and pH 7. Waterbinding and firmness obviously do not correlate, since a minimum in moisture loss was observed at 75–78°C, whereas the firmness increased with heating temperature. When evaluating protein ingredients for a given product, it may be useful to identify which of these properties contributes most to the quality of the final product.

Some aspects of methodology

The discrepancy between the waterbinding and texture characteristics is of importance for the choice of methodology. When the ability of the gel to bind water is to be determined, conditions have to be chosen so the influence of the gel texture is minimized.

The term waterbinding is used here as a general term and does not refer to the small portion of physically bound water with properties significantly different from those of free or bulk water. Common methods to determine waterbinding are based on the application of an external force or on capillary suction of a porous material in contact with the sample [3, 6, 7, 8, 9, 10].

On application of an external force, the waterbinding is either expressed as the waterholding capacity (WHC), which is the amount of water bound per gram protein, or as the reverse moisture loss (ML), which is the amount of juice released per gram of gel sample.

A dangerous test for three-dimensional structures is centrifugation at

Table 1. The effect of experimental conditions on percent moisture loss values of 5% plasma protein gels at pH 9.0, heated at 82°C, in the high speed centrifugation test [3]

Centrifugation force (X g)	Small centrifuge tubes (Ø 10 mm) sample weight (g)				Big centrifuge tubes (Ø 27 mm) sample weight (g)			
	1	2	3	5	5	10	15	20
5090	6.9	3.8	2.1	1.1	20.0	13.2	9.0	6.7
9750	13.1	6.6	4.2	2.0	28.1	19.3	13.0	9.0
20200	–	–	–	–	34.1	29.8	22.7	18.0
30000	41.1	38.3	37.2	36.9	51.1	43.3	34.9	31.9
39100	–	–	–	–	55.3	50.2	46.4	45.8

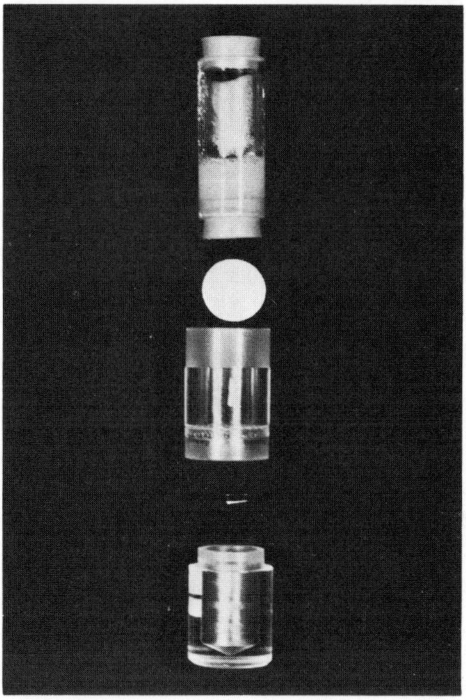

Figure 6. Apparatus for conducting modified net test [3].

high speed and measurement of the juice released above the sample in a centifugation tube. In such a test the deformability determines the amount of water released. If the structure is weak it may break down during centrifugation and a measure of subfractions is obtained rather than the structure under investigation [3, 10]. The data in Table 1 show the effect of sample size and centrifugation force on the moisture loss of identically prepared gels. At high speeds, the gels collapsed and the results are highly dependent on sample geometry. This method is obviously not reliable for the determination of waterbinding properties of gels.

A test that seems to work reasonably well for gel systems is a modification

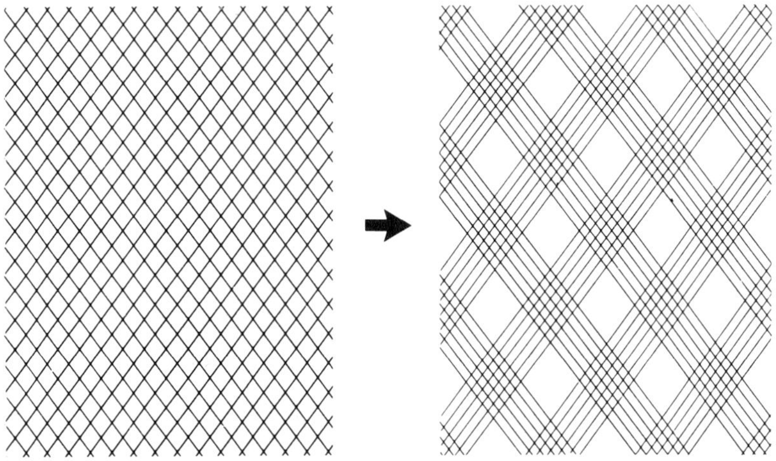

Figure 7. Schematic illustration of a change in gel structure due to local aggregation phenomena [5].

of the so called 'net-test' [3]. The principle was first described by Wierbecki and used for studies of meat systems [6]. The modified set-up used in our laboratory for protein gels as well as complex food systems, is shown in Figure 6.

The gel is formed in the upper plexiglass tube. After cooling the bottom rubber stopper is removed and the tube is attached to the middle section. This section has a net in the bottom to allow drainage of water to the bottom section. For work with protein gels a filter paper is placed on the top of the net. The whole assemblage is put in a centrifuge tube and centrifuged very gently so the gel sample is not deformed. Moisture loss is determined by weighing before and after centrifugation. Waterholding capacity can be calculated by determining the protein or dry content of the gel plug after centrifugation. The drip can also be collected and analyzed.

It should be pointed out that this is also a relative test that does not hold for all types of gels. The perfect test for determining waterbinding properties of gels has not yet been developed. So far we consider that the modified net test is the best one available for practical work.

Relationships between microstructure and functional properties

The discrepancy in waterbinding and texture properties of plasma protein gels demands an explanation. A hypothesis can be put forward to explain how a change in the gel structure could give rise to a decrease in waterbinding at the same time as the firmness increased.

A gel can be regarded as an intermediate state between a solution and a precipitate with just the right balance between protein–protein and protein–solvent interactions. If heating above the gelation temperature increases the

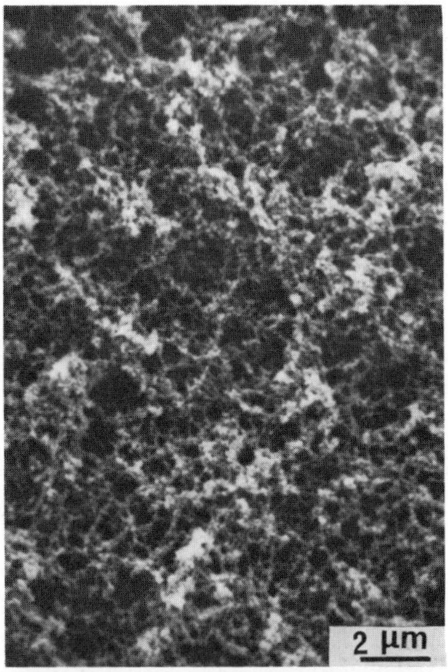

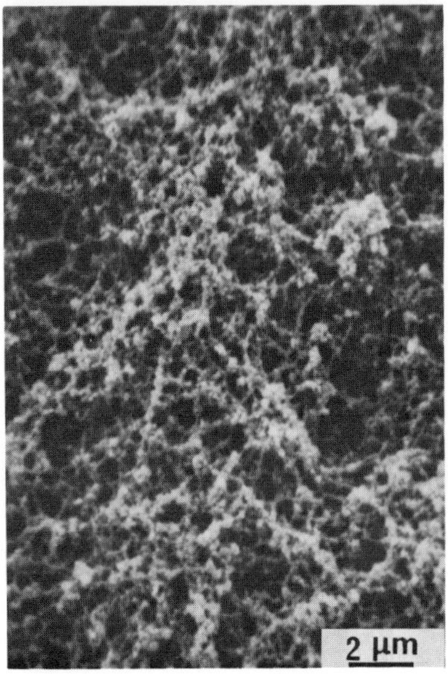

Figure 8 (upper left). SEM micrograph of 5% plasma protein gel made at pH 9.0 and heated to 77°C.

Figure 9 (upper right). SEM micrograph of a 5% plasma protein gel made at pH 9.0 and heated to 92°C.

Figure 10 (bottom left). SEM micrograph of a 5% plasma protein gel made at pH 7.0 and heated to 92°C.

Figure 11 (bottom right). SEM micrograph of a 5% plasma protein gel made at pH 7.0 and heated to 92°C.

Table 2. Generalized outline of the effect of conditions during gel formation on aggregation, waterbinding, elasticity and firmness[a]

Variables[b]	Range	Aggregation	Waterbinding	Elasticity	Firmness[c]
Heating temperature at pH 7 and 9	77 to 92°C	+	—	—	+
pH	pH 6 to 9	—	+	+	+
Protein conc. at pH 9	4 to 6%	+	—	—	+
Protein conc. at pH 7	4 to 6%	(+)	—	±0	+
NaCl conc. at pH 9	0 to 5%	+	—	—	+
NaCl conc. at pH 7	0 to 5%	(+)	±0	±0	±0

[a]From [5].
[b]Except as otherwise indicated, 5% plasma protein gels were formed at 82°C without salt addition.
[c]Firmness as measured by compression or penetration tests in an Instron Universal Machine.

tendency for protein–protein interactions, either of two events can take place. Shrinkage will occur if protein–protein interactions are uniform throughout the sample. If not, locally stronger interactions will cause a partial disruption of the gel network, and the gel will become more aggregated, as schematically illustrated in Figure 7.

Such a change gives a possible explanation of the increase in moisture loss, since water is easily pressed out through the larger capillaries in the structure. During compression, the increased amount of denser regions may contribute to an increase in the registered force.

Studies of the microstructure of blood plasma gels by scanning electron microscopy (SEM) revealed that the postulated changes really took place. Figure 8 and 9 show gel structures of 5% plasma protein gels at pH 9, heated to 77°C and 92°C, respectively. Figure 10 and 11 show the corresponding micrographs of gels made at pH 7.

At both pH values, the gel heated to 92°C is coarser and more aggregated than the gel formed at 77°C. The gel formed at pH 7 is coarser and more randomly aggregated than the gel formed at pH 9 at both temperatures due to the lower negative net charge at pH 7 (compare Figure 8 and 10; Figure 9 and 11). It is interesting to note that gels of similar work density were observed when the gel at pH 9 was heated to 92°C and the gel at pH 7 to 77°C. This indicates the possibilities of controlling the structure formation by a proper knowledge of the mechanisms involved.

Plasma gels prepared under a wide range of conditions such as different pH, heating temperature and salt and protein concentration, were investigated. It was found that changes known to enhance the tendency towards random aggregation always decreased the waterbinding properties of the gels. Changes in texture parameters related to firmness were more complex and reacted

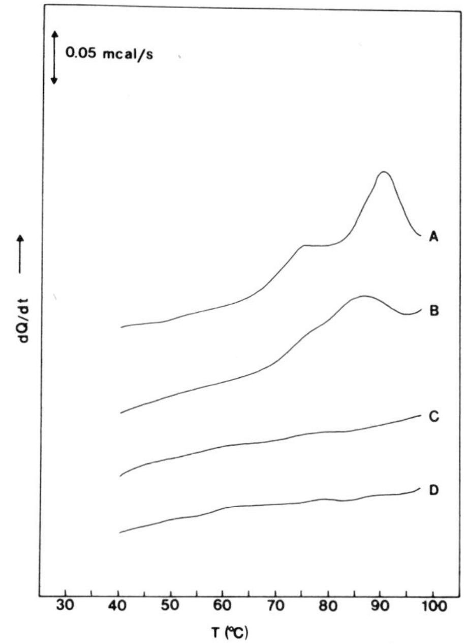

Figure 12. DSC thermograms of 10% soy protein isolates at pH 7.0 in distilled water. Isolate A produced under mild conditions. Isolates B, C, and D commercially available isolates. The sensitivity was 0.5 mcal/s and the heating rate 10°C/min [2].

differently to different variables. The reason for this is that firmness, when measured as the compression force, reflects a sum of all events taking place during deformation, and it is not yet possible to predict the contribution of various structure elements to the overall texture. In four out of five cases, firmness was positively correlated to aggregation. However, when a measure of the degree of elasticity was studied instead of firmness, it was found to be negatively correlated to the degree of random aggregation in all cases. Thus, different texture parameters give different kinds of information. The results are summarized in Table 2 [3, 4, 5].

Effects of processing on structure and functional properties

Vegetable proteins were not included in the general discussion on relationships between structure and functional properties, since they are complex and often modified by processing. Thus commercial soy protein isolates differ substantially from mildly prepared isolates in many respects, and it is dangerous to extrapolate results obtained from one type of soy protein to another.

Processing may cause a partial or complete denaturation, as shown by the DSC thermograms in Figure 12. Curve A represents the thermogram of a mildly produced isolate with the two characteristic peaks of the 7S and 11S globulins. Isolates B, C, and D are commercially produced isolates, where

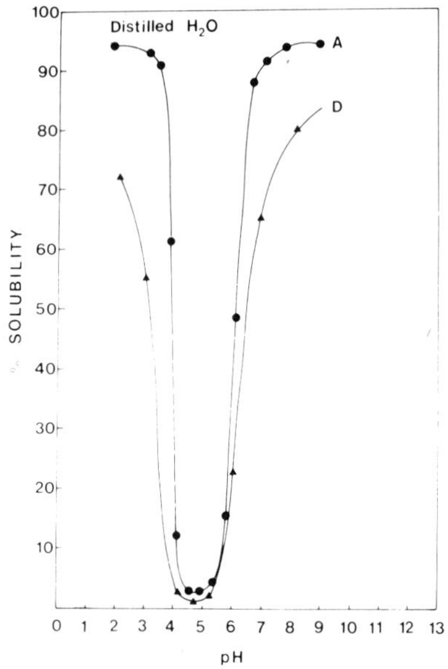

Figure 13. Solubility as a function of pH for 1% dispersions of the mildly produced soy protein isolate A and the commercial soy protein isolate D in distilled water [2].

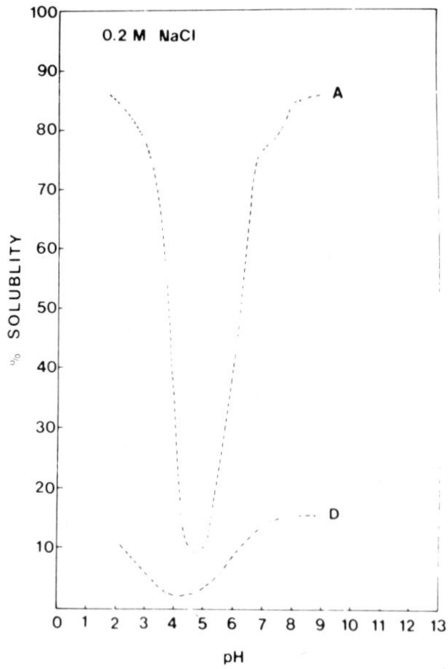

Figure 14. Solubility as a function of pH for 1% dispersions of the mildly produced soy protein isolate A and the commercial soy protein isolate D in 0.2 M NaCl [2].

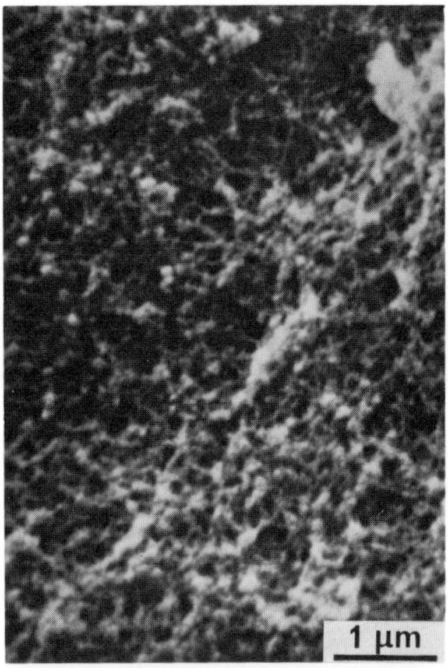

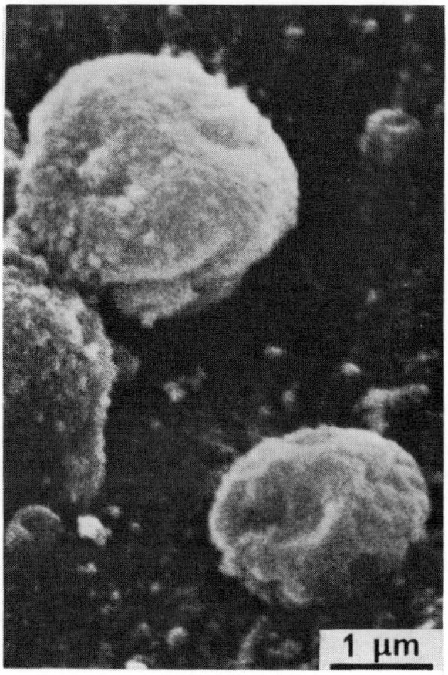

Figure 15 (upper left). SEM micrograph of a 12% soy protein gel of isolate A made in distilled water [12].

Figure 16 (upper right) SEM micrograph of a 12% soy protein gel of isolate A made in 0.2 M NaCl [12].

Figure 17 (bottom left). SEM micrograph of a 12% soy protein gel of isolate D made in distilled water.

Figure 18 (bottom right) SEM micrograph of a 12% soy protein 'gel' of isolate D made in 0.2 M NaCl.

isolate B has been partly denatured and isolates C and D completely denatured during processing.

The degree of denaturation has an impact on functional properties under certain conditions. When the solubility was measured as a function of pH, a dramatic difference in the solubility profile between isolates A and D was found in the presence of 0.2*M* NaCl, but not when the measurements were made in distilled water (see Figures 13 and 14). The low solubility of isolate D in the presence of salt is of practical importance, since no foods are salt-free. Meat products which are an important field of application for soy products contain 0.2 to 1.0*M* NaCl [11].

The behaviour of soy proteins in the presence and absence of salt was further investigated in a model study of gels. Addition of salt has previously been shown to favour protein–protein interactions and to cause gels of mildly prepared soy proteins to have a higher degree of random aggregation than gels made in distilled water [12]. Examples from this study are given in Figures 15 and 16. SEM micrographs at a magnification of 12000 are shown for comparison, even though details at the molecular level are better seen from TEM (Transmission EM) studies at higher modifications. The corresponding SEM micrographs of the commercially produced isolate D are shown in Figure 17 and 18. All gels were made at 12% protein concentration and heated at 85°C for 30 min. The gel of isolate D is of the same dimension as the gels of isolate A, when made in distilled water. In the presence of salt only two particles are seen. At a much lower magnification of 650 × it can be seen that the structure is built entirely of swelled but undissolved particles (see Figure 19). The particles shown in Figure 18 are encircled and it is seen that they are small relative to the particles making up the main structure. The limited swelling of isolate D makes it easy to identify in cooked frankfurters, as illustrated in Figure 20.

The results obtained in simple model systems on solubility and gel characteristics indicate the possibility of manipulating the function of commercial soy proteins in meat products by dissolving them in pure water before adding them to the meat mixture.

Aspects of protein concentrates

Gluten is another important vegetable protein. It is vital to the function of bread and there is an increasing interest in its use in other fields of application. Commercially available gluten products can be considered as concentrates containing 72–76% protein (N × 5.7), 5–11% starch, 2–8% fibres, 5–7% fat and 4–10% water [13].

If gluten is considered as an ingredient of non-bread products, other properties than dough and film characteristics may be of importance. A study was therefore made of the effect of temperatue on waterbinding, texture and microstructure of various gluten products [13]. Some results from this study were chosen to illustrate the importance of other components than protein for the functional properties of a concentrate.

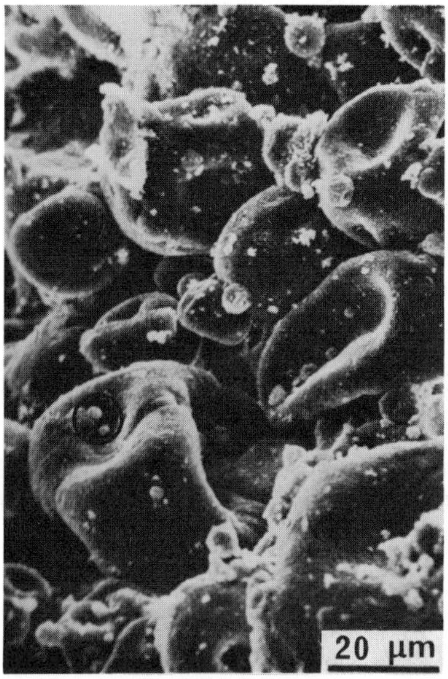

Fig. 19

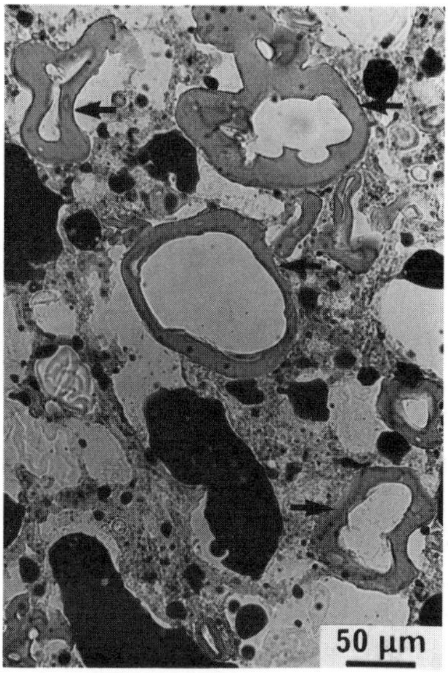

Fig. 20

Figure 19. SEM micrograph of a 12% soy protein 'gel' of isolate D made in 0.2 M NaCl at a lower magnification. The particles shown in Figure 18 are encircled.
Figure 20. Light micrograph of a cooked sausage where swelled particles of isolate D are easily identified. Some particles are marked with arrows.

Figures 21A–D show various function characteristics of a hydrated gluten product as a function of heating temperature. From Figure 21A, it can be seen that the product expanded on heating to temperatures above 80°C. The variations in waterholding capacity (WHC) were relatively small as shown in Figure 21B. WHC increased somewhat between 50 and 80°C and there was a maximum after storage at − 5°C, probably due to a certain amount of freeze texturization at this low freezing temperature. The texture changes were more dramatic, as can be seen in Figures 21C and D. There was a marked increase in penetration force at temperatures above 80°C. The degree of elasticity, when measured as the time required for 50% force relaxation, had already increased at temperatures between 55°C and 60°C. The gluten gels were made in open end plexiglass tubes such as the one shown in Figure 6.

Possible causes of the temperature-induced changes were studied by DSC and microstructure evaluation. Figure 22 shows DSC thermograms of gluten, as well as the corresponding thermograms of wheat meal, and A- and B-starch fractions. The thermograms show that the only endothermic phase transition taking place is due to changes of the starch structure. Gluten prepared by hand washing gave the same flat DSC curve as the commercially produced gluten product.

Figure 23 shows SEM-micrographs of hydrated gluten treated at 25°, 75°, 95°, 121°C, and − 15°C followed by critical point drying. At 25°C the B-starch granule was embedded in a dense protein matrix. At 75°C the starch

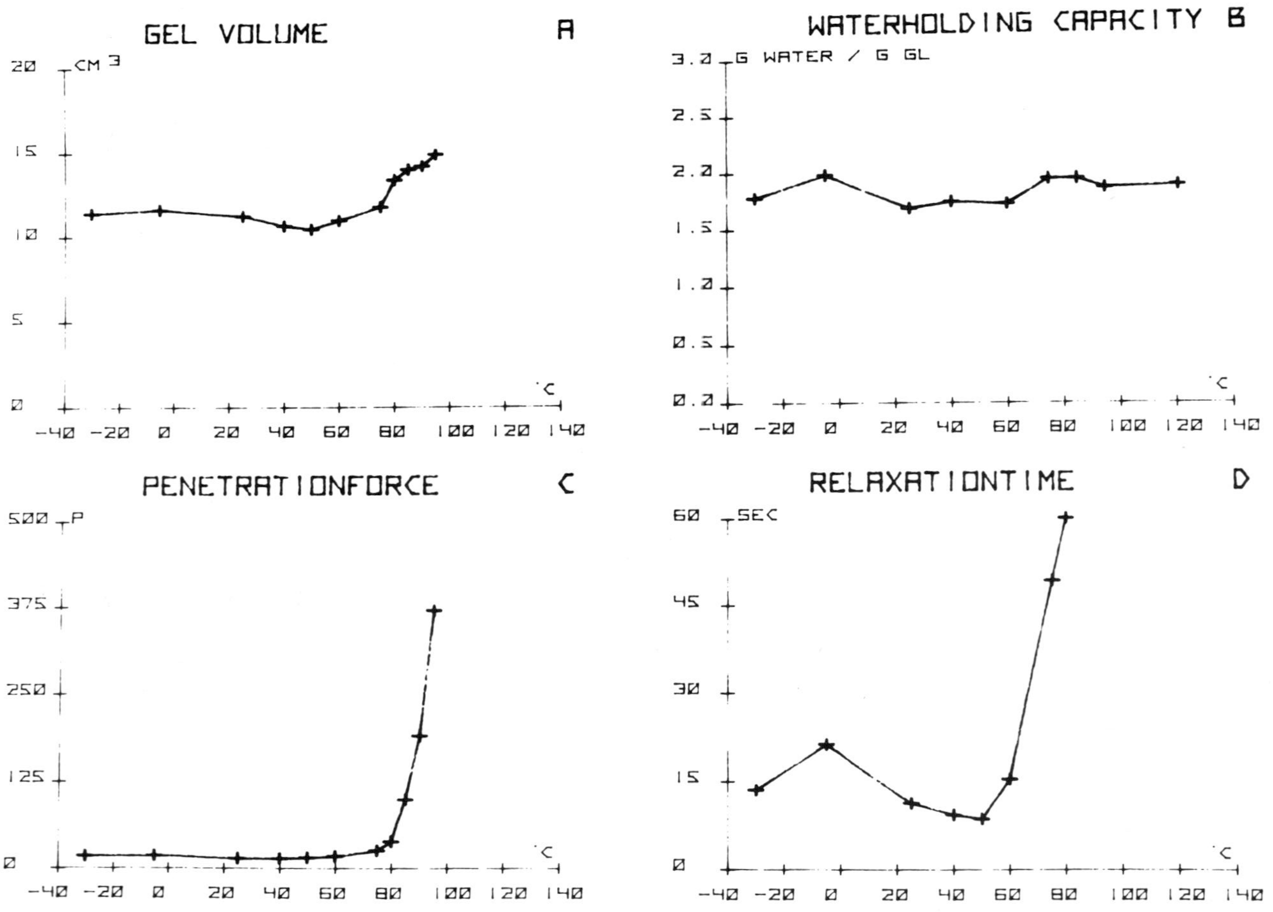

Figure 21. Gluten properties as a function of heating temperature. All samples were hydrated before treatment at a gluten to water ratio of 1 to 3. A. Volume change in open end plexiglass tubes. B. Waterholding capacity determined according to the modified net test. C. Penetration force at a penetration depth of 10 mm with 5 mm plunger. D. Elasticity measured as the time required for 50% force relaxation from a compression force of 1 kp

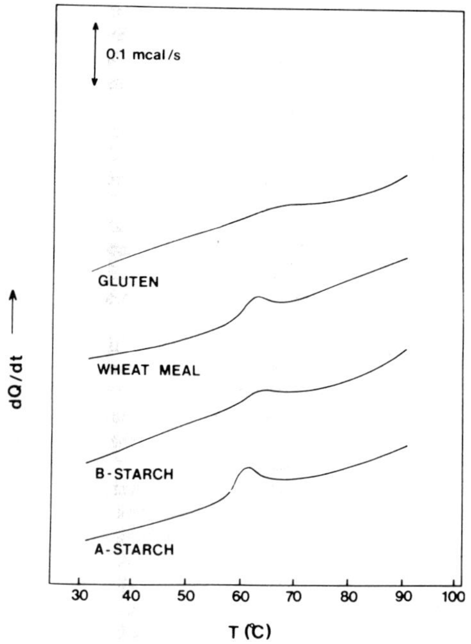

Figure 22. DSC thermograms of hydrated gluten, wheat meal, A- and B-starch. The sensitivity was1.0 mcal/s and the heating rate 10°C/min. [13].

granule has swelled and part of its content has leaked out but the protein matrix has not changed. At 95°C and 121°C the identity of the granule is lost and remains of swelled starch are seen in the depression made by the granule. The structure of the protein matrix looks the same as at 25°C. The micrographs taken after freeze storage at −15°C, thawing and preparation by critical point drying look almost identical to that of the untreated gluten at 25°C. The very dense protein structure found by critical point drying was confirmed by special freeze etching techniques and TEM studies at a higher magnification.

A large number of SEM-micrographs presented in the literature on gluten and gluten protein fractions show a much coarser structure than the very dense protein matrix discussed above. This is most probably due to artifacts caused by freeze drying. Ice crystals form on freezing and distort the structure. If the structure is fixed by drying from the frozen stage, an artifactural structure will be studied. This is illustrated at the bottom right of Figure 23 showing the structure of the same sample as that shown in the other micrographs of gluten but prepared by freezing in liquid nitrogen, followed by freeze drying. A regular network caused by ice crystal formation is seen. If thawed, the sample recovered and a similar structure to that shown at the bottom left was obtained. Under special conditions irreversible freeze texturization can take place, as indicated by Figure 21B.

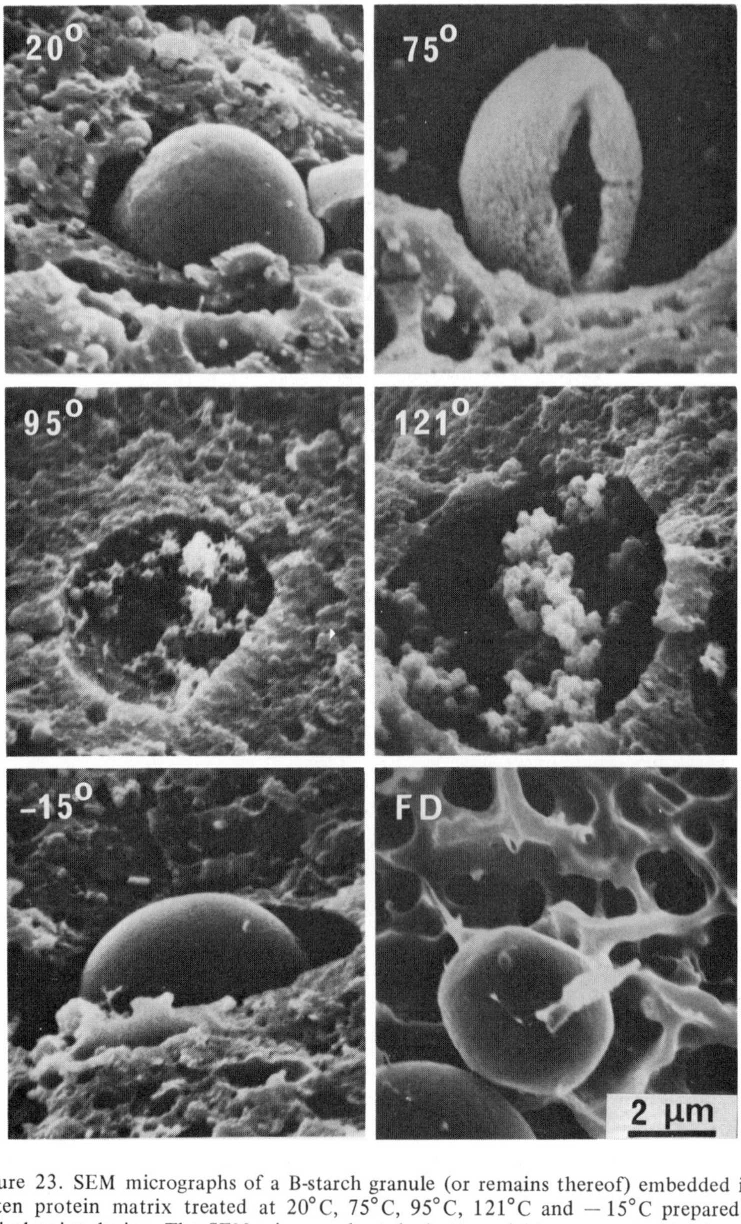

Figure 23. SEM micrographs of a B-starch granule (or remains thereof) embedded in a gluten protein matrix treated at 20°C, 75°C, 95°C, 121°C and −15°C prepared by critical point drying. The SEM micrograph at the bottom right was prepared by freezing in liquid nitrogen followed by freeze drying. The gluten to water ratio was 1 to 3 [13].

The micrographs shown in Figure 23, together with the DSC curves, clearly show that the major changes taking place during heat treatment of hydrated gluten are due to starch swelling and not to changes in the protein. A starch content of 10% by weight may not seem a lot, but the B-starch

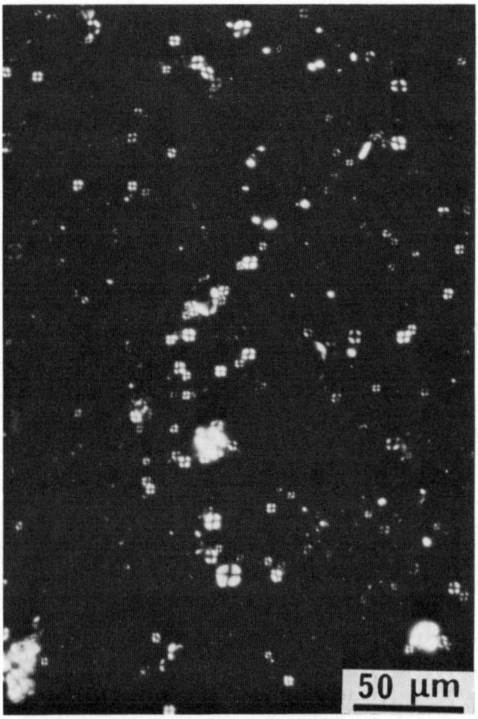

Figure 24. Light micrograph taken in planepolarized light of a hydrated gluten section showing the distribution of birefringent starch granules. [13]

granules mainly included in the gluten protein matrix are small and already make up a lot of the area to volume ratio in the unswelled state, as shown in Figure 24.

Conclusions

By the results shown, an attempt has been made to illustrate various aspects which are important for functional characterization. Properties to be determined have to be better defined from the physical point of view, and their relevance for the function of an ingredient in a food product well-established. The field of functional characterization lies within the scientific gap between the basic science of well-defined systems which do not necessarily give information relevant to complex systems, and applied work on food products, where cause and effects are seldom known. For future work it is important to deepen our understanding of functional properties and their relationships to the microstructure and composition of food ingredients.

Acknowledgements. Thanks are due to M. Langton and G. Schallin for preparation of electron micrographs and to A. Granquist and E. Hast for analysis of functional properties.

References

1. Hermansson, A-M (1972) Determination of functional properties of protein foods, chapter 27. In: Porter and Rolls, (eds), Proteins in Human Nutrition AP.
2. Hermansson, A-M (1979) Methods of studying functional characteristics of vegetable proteins. J Amer Oil Chem Soc 56(3):272–279
3. Hermansson, A-M and Lucisano M (1982) Gel characteristics – waterbinding properties of blood plasma gels and methodological aspects on waterbinding of gel system. J Food Sci 47(6):1955–1959 & 1964
4. Hermansson, A-M (1982) Gel characteristics – compression and penetration of blood plasma gels. J Food Sci 47(6):1960–1964
5. Hermansson, A-M (1982) Gel characteristics – structure as related to texture and waterbinding of blood plasma gels. J Food Sci 47(6):1965–1972
6. Wierbecki E, Kunkle LE and Deatherage FE (1957) Changes in the waterholding capacity and cationic shifts during heating and freezing and thawing of meat as revealed by a simple centrifugation methods for measuring shrinkage. Food Technol 11(2):69
7. Hermansson A-M and Åkesson CÅ (1975) Functional properties of added proteins correlated with properties of meat systems. Effect of concentration and temperature on waterbinding properties of model meat systems. J Food Sci 40:595–602
8. Oshiai- Yanagi S, Miyauchi H, Saio K and Watanabe T (1978) Modified soybean protein with high waterholding capacity. Cereal Chem 55(2):157–167
9. Hofmann K (1975) Ein Neues Gerät zur Bestimmung des Wasserbindung des Fleisches. Das 'Kapillar-Volumeter'. Fleischwirtschaft 55, H1:25–30
10. Labuza TP and Lewicki PP (1978) Measurement of gel waterbinding capacity by capillary suction potential. J Food Sci 43:1264–1273
11. Hermansson A-M and Åkesson CÅ (1975) Functional properties of added proteins correlated with properties of meat systems. Effect of salt on waterbinding properties of salt on waterbinding properties of model meat systems. J Food Sci 40:603–610
12. Hermansson A-M and Buchheim W (1981) Characterization of protein gels by scanning and transmission electron microscopy. A methodology study of soy protein gels. J Coll Int Sci 81(2):519–530
13. Hermansson A-M (1980) Project No. P-226, of Nordisk Industrifond. Gluten Study at SIK supported by Nordisk Industrifond, Raision Tehtaat and Alfa-Laval

Comparative physical and chemical aspects of vegetable protein functionality

DAVID J. WRIGHT

A.R.C. Food Research Institute, Division of Chemistry and Biochemistry, Colney Lane, Norwich NR4 7UA, England

Key words: legumin, glycinin, physico-chemical properties, structure, functional properties

Abstract. Economic and political reasons may make it advantageous to be able to substitute one source of vegetable protein for another. Just how feasible this proposition is will depend to a large extent on the degree of similarity exhibited in the properties of these various proteins. The latter are utilised in food systems for their functional properties, conferring desired processing behaviour or product attributes. These functional properties must in some way be related to the physico-chemical properties of the proteins, which in turn must be a function of their inherent composition and structure. For example, the physico-chemical properties of a protein which might conceivably contribute to a functional property such as solubility, are hydrophobicity, net surface charge and molecular weight. At a more fundamental level these properties can be regarded as functions of the amino acid composition, primary sequence and structure (conformation). Similarities in these basic characteristics of proteins can, therefore, be assumed to imply a similarity in both their physico-chemical and functional properties. To what extent this statement applies to vegetable proteins is considered in this discussion. Soyabean, field bean, and pea are taken as the principle examples and certain aspects of their structural, physico-chemical and functional properties compared.

Introduction

Currently, the vegetable protein field is dominated by soyabean protein, in one form or another. However, economic and political factors may make it advantageous, if not desirable, to be able to substitute for soya protein other vegetable proteins derived from other sources, in particular those crops more suited to the climatic and agronomic conditions of the country concerned. The feasibility of such a proposition will depend to a large extent on the degree of similarity exhibited by these proteins in their various properties.

Soya protein is utilised in food systems for its functional properties which confer desired processing behaviour or, more importantly, essential product attributes. There is a wide range of functional properties ascribed to proteins in general and some of the more important ones are listed in Table 1. These properties fall into three broader categories: (1) hydration, (2) absorption, and (3) texture/structure generation or stabilisation, reflecting the different

Table 1. Aspects of functional, physico-chemical and structural properties of proteins

Functional properties	Physico-chemical properties	Structural properties
Solubility	Surface charge	Amino acid composition
Gelation	Molecular weight	Amino acid sequence
Emulsifying capacity	Hydrophobicity	Secondary, tertiary and quaternary structure
Foaming ability	Association/dissociation behaviour	
Fat/water binding	Thermal stability	

applications of proteins in food systems. The functional properties must be a reflection of the physico-chemical properties of the proteins and a selection of these are shown in Table 1. In their turn, the physico-chemical properties of a protein must be a manifestation of the protein's inherent composition and structure, in other words, its amino acid composition, the arrangement of amino acids within the protein (primary sequence) and finally the organisation of polypeptide chains and sub-units in the native protein (secondary, tertiary and quaternary structure). The first two are independent variables since it is obviously possible to synthesise innumerable amino acid sequences from a pool of amino acids. However, this is not true for the higher structural elements which present evidence suggests are largely determined by the primary amino acid sequence of the protein. Nevertheless, for the time being, it is necessary to include these parameters because the exact relationship between them and the amino acid sequence is not yet known. Unfortunately, for only a very few proteins is the actual three-dimensional structure known: this must remain an important objective if a thorough understanding of the properties and behaviour of any protein is to be gained. In the meantime, physical techniques can provide us with data on certain aspects of these higher levels of protein organisation such as content of α-helix, β-pleated sheet, and random coil configuration as well as sub-unit composition.

The precise relationship between the structure of a protein and its functional properties is not known and so food scientists have endeavoured to relate the latter to the expression of the protein's structure as measured by a range of physico-chemical parameters. For example, the size as given by the molecular weight, the charge and polarity as given by the isoelectric point and average hydrophobicity, and finally the intramolecular bonding forces as given, for example, by the dissociation behaviour and stability to denaturation. At an intuitive level it is possible to select those physico-chemical properties which most obviously contribute to specific functional properties. For example, a functional property such as solubility might be influenced by the molecular weight, hydrophobicity and net surface charge of the protein. Because, at a more fundamental level, these latter properties can all be regarded as functions of the amino acid composition, primary sequence and structure of the protein, it then follows that similarities in

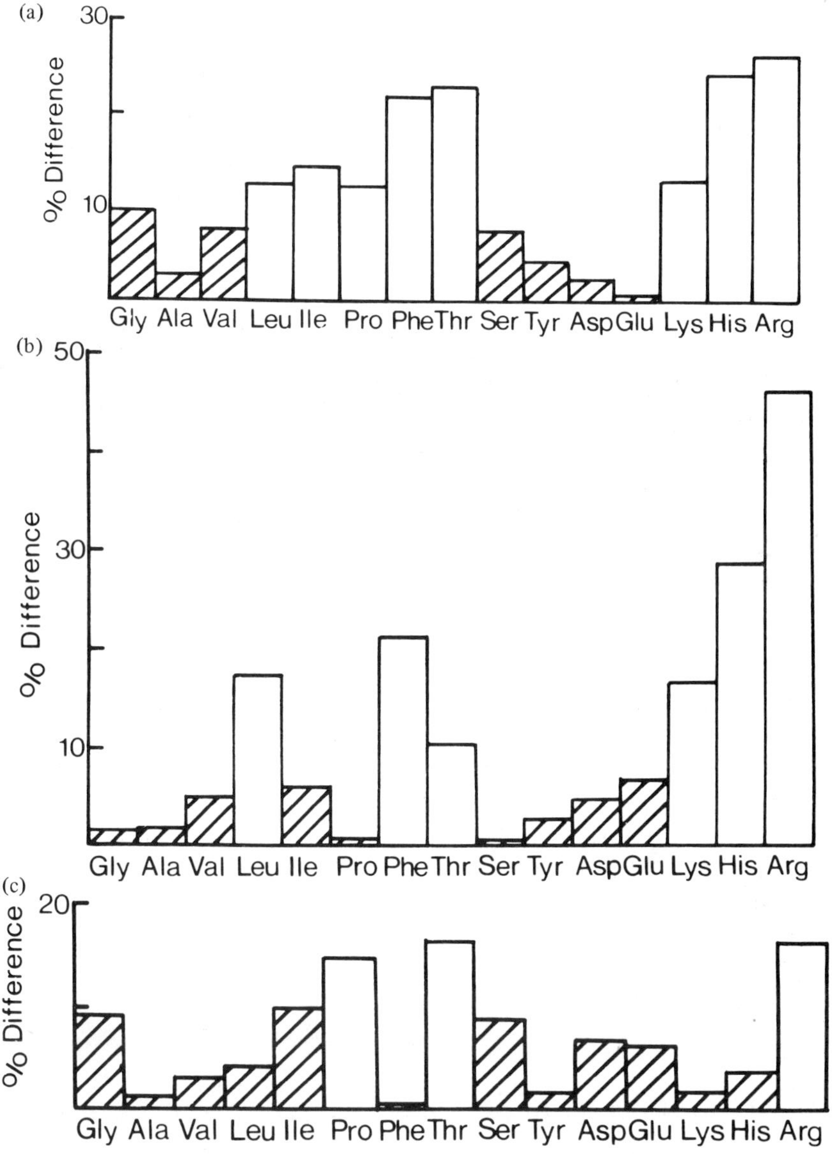

Figure 1. Difference amino acid profiles of (a) legumin (pea) vs. glycinin (soya), (b) legumin (field bean) vs. glycinin (soya), and (c) legumin (pea) vs. legumin (field bean). Hatching indicates those amino acids differing by 10% or less between the two proteins. Data taken from Derbyshire et al. [1], Mori and Utsumi [2], Casey [3], and Badley et al. [4].

these structural characteristics between proteins can be assumed to imply a corresponding similarity in both their physico-chemical and functional properties. To what extent this assumption is valid for vegetable proteins will now be discussed using soyabean, field bean and pea as the examples.

Because of the lack of comparable data for the 7S globulins of these legumes, only the 11S or legumin-like storage proteins will be considered here.

Structural characteristics of the 11S globulins

Amino acid composition

A direct comparison of the amino acid profiles of these three proteins indicates a gross similarity between them [1]. In order to accentuate these similarities and also any differences, the amino acid data is presented in a modified form in Figure 1. Here the percentage differences between one protein and another in the content of fifteen of their constituent amino acids are shown. Where differences of 10% or less occur, these are indicated by hatching. Those amino acids which are present at very low concentrations (e.g. cysteine) are excluded from this analysis because of the uncertainty and variation in the published data. In Figure 1(a) legumin from pea is compared to glycinin from soya. For these two proteins, a total of seven amino acids fall within the 10% difference limit with none of the remainder displaying a divergence greater than 20%. Some points to note are that the small non-polar amino acids (alanine, valine, glycine) and the dicarboxylic acids/amides (glutamic, aspartic) are conserved whilst the three basic amino acids (lysine, histidine, arginine) show appreciable variations. Important structural information can be gleaned from such data, in the sense that amino acids which are consistently conserved between like proteins may reflect the presence of essential structural elements or features in the proteins.

Comparison of legumin from *Vicia faba* with glycinin (Figure 1(b)) reveals nine amino acids which differ by less than 10%. Once again it is the small non-polar and the dicarboxylic amino acids that are conserved.

From the preceding two figures it is fairly obvious that if the two legumin proteins are compared, the similarity will be even more striking. This is indeed the case, as Figure 1(c) shows. Only three of the amino acids, arginine, threonine and proline, differ by more than 10% and even in these cases the differences are less than 20%.

Of course it could be argued that the differences apparent in the amino acid content of these proteins is large for supposedly similar proteins. This, however, might merely be a reflection of their role as storage proteins, which, as such, are presumably subject to less stringent evolutionary controls than a protein with, for example, enzymatic functions. Nevertheless, it is instructive to repeat this exercise with a protein of entirely different function, such as the haemagglutinin in soya, to assess the significance of this variation. When the amino acid profile of soya haemagglutinin is compared to that of glycinin (Figure 2) the differences are clearly much greater than those observed between the 11S storage proteins. All amino acids differ by greater than 10%, most by more than 20% and some by as much as 50–60% (arginine and

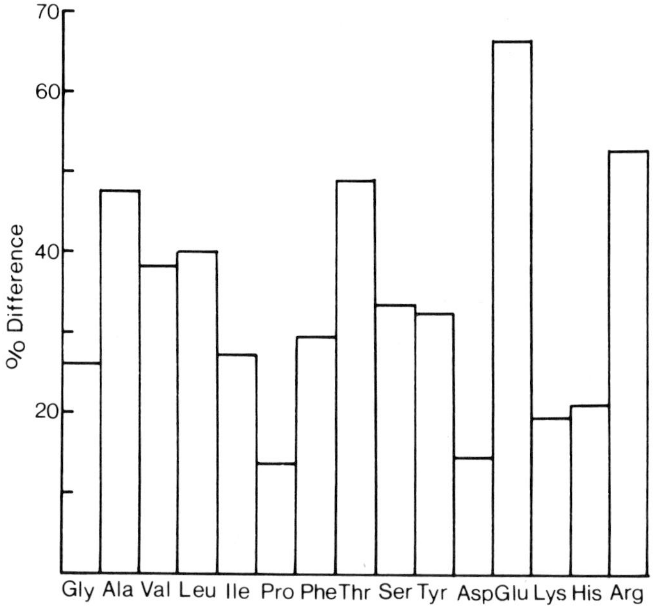

Figure 2. Difference amino acid profile of haemagglutinin (soya) vs. glycinin (soya). Data taken from Derbyshire et al. [1], Badley et al. [4] and Lotan et al. [5].

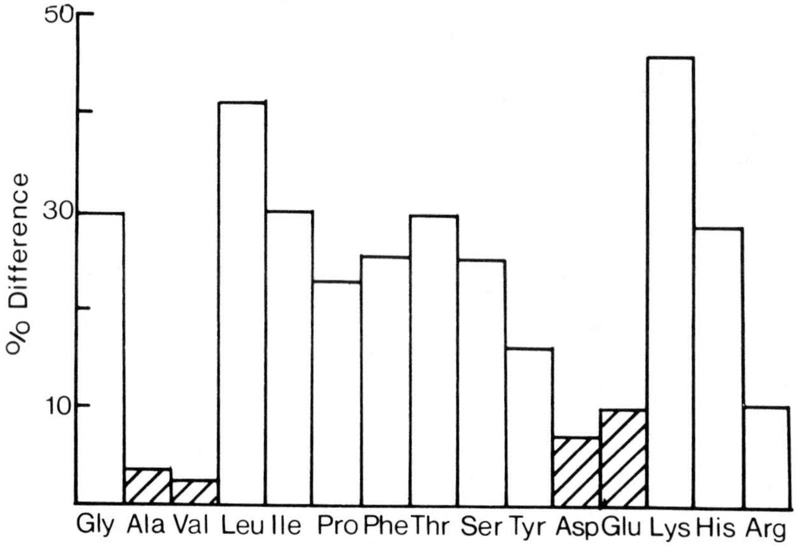

Figure 3. Difference amino acid profile of β-conglycinin (soya) vs. glycinin (soya). Hatching indicates those amino acids differing by 10% or less between the two proteins. Data taken from Derbyshire et al. [1], Badley et al. [4], and Koshiyama [6].

glutamic acid). Thus the difference in function is directly reflected in their amino acid composition.

This approach can be extended one stage further by comparing the amino acid compositions of proteins with the same storage function but of different structure, i.e. the 7S and 11S storage proteins. A comparison of β-conglycinin and glycinin,both from soyabean, reveals a total of five amino acids where the difference is less than 10% and once again it is the two non-polar amino acids, alanine and valine, and the two dicarboxylic acids (amides), glutamic and aspartic that are conserved (Figure 3). Clearly these four amino acids must play an important role in the structural organisation of storage proteins. The two non-polar amino acids may be required for close-packing of residues in the hydrophobic core of the proteins, whilst glutamic and aspartic acids and their amides may be important contributors to the stabilisation of the proteins' structure via hydrogen bonding.

Of course, ultimately, the real significance of these similarities in amino acid composition can only be assessed in relation to the ordering of these amino acids in the protein, i.e. the primary structure.

Primary structure

Although some progress in the sequencing of vegetable proteins has, and is, being made, the enormity of the task has meant that only a few partial sequences are known so far [7–9]. For comparative studies, the available information is restricted to the N-terminal sequences of the basic sub-units of the 11S globulins. Although the latter are known to be heterogeneous with respect to both size and charge [10–12], the data presented here (Figure 4) is of total basic sub-units rather than of separated individual sub-units. Despite this, there is seen to be a remarkable correspondence in amino acid sequence between these sub-units from the three proteins, the homologous sections being indicated by the dashed line beneath. This similarity appears even greater if only pairs of the proteins are considered or if the comparison is made between individual basic sub-units of the three proteins, i.e. if allowance is made for the presence of common residues at positions where more than one amino acid is observed (e.g. positions 5, 6, 13 and 21). Furthermore, certain amino acid substitutions are unlikely to affect dramatically the structure of these proteins, for example, the replacement of one non-polar residue with another at positions 2 and 6, and the substitution of one dicarboxylic acid by the other at position 3. One other interesting feature of these sequences is the conservation of the amino acid cysteine at position 7, indicating perhaps that it may be involved in the disulphide bond linking acidic and basic sub-units of these proteins.

Higher order structure

This, the last of the structural properties mentioned earlier, relates to the protein conformation and sub-unit composition. Some information on the conformation of the 11S globulin from soya is published [13, 14], but

N-TERMINAL AMINO ACID SEQUENCE OF β-SUBUNITS OF 11S PROTEINS

	1				5					10					15
Vicia faba :	G	L	E	E	T	V/I	C	T	V/A	K	L	R	L/E	N	I
Pisum sativum :	G	L	E	E	T	V/I	C	T	A	K	L	R	L/E	N	I
Glycine max :	G	V/I	D	E	N/T	I	C	T	M	K	L	R	E	N	I

					20					25					
Vicia faba :	A/G	‹Q›	P	‹A›	R	P/A	D	L	Y	N	P	‹Q›	A	G	
Pisum sativum :	A/G	P	‹S›	‹S›	‹A›	P	D	L	Y	N	P	E	A	G	
Glycine max :	A/G	Q	P	‹S›	R	P/A	D	L	Y	N	P	‹Q›	A	G	

Figure 4. The N-terminal amno acid sequence of the basic sub-units of 11S globulins from soya, pea and field bean. Residues in parentheses indicates some uncertainty in the identification. Data taken from Casey et al. [9] and Gilroy et al. [7].

the only comparable data available for all three proteins concerns their sub-unit structure. Each one has been shown to consist of 6 sub-units which in turn are comprised of one acidic and one basic sub-unit joined by di-sulphide bonds [10, 15, 16].

Thus, overall, from the point of view of structural properties, these three storage proteins, although by no means identical, are clearly very closely related. The next problem is to see whether this similarity is also expressed in their physico-chemical properties.

Physico-chemical properties

Table 2 shows a selection of physico-chemical properties reflecting three major characteristics of the proteins: structure, viz., size (and polarity), charge and internal bonding. The sizes, as given by the molecular weights, correspond fairly well allowing for experimental error in their determination. (It should be emphasised, however, that the higher molecular weight of legumin from *Pisum sativum* is real, in the sense that it is consistent with the slightly higher molecular weights of its sub-units [12].) With regard to surface charge, the isoelectric points of the three proteins would indicate that they all possess a similar balance of acidic and basic charged groups on the surface of the molecule. In addition, the computed values for the average hydrophobicity of these proteins suggest a correspondence in their inherent polarity. The actual values are indicative of fairly moderate hydrophobic tendencies and this is consistent with the observed solubility properties of these proteins in water or dilute salt solution.

Table 2. Some physico-chemical properties of 11S globulins

11S globulin	Mol. wt. (daltons) [4, 10, 17]	pI [14, 18, 19]	Average hydrophobicity[a] (Kcal residue^{-1})	Dissociation behaviour [1]
Glycinin (*G. max*)	320 000	4.64	0.94	11S → 7S → 3S → 2S
Legumin (*V. faba*)	330 000	4.8–5.5	0.96	11S → (?) → 3S → 2S
Legumin (*P. sativum*)	350 000	4.8–5.0	0.93	11S → (?) → 3S → 2S

[a]Calculated from the amino acid composition and side chain hydrophobicities [20].

The question of the nature of the intramolecular bonds present in the molecule is an important one because this aspect of the protein's structure will dictate the stability towards unfolding or denaturation, and the latter is central to such functional properties as gelation, emulsification and foaming. Some information on the essence of the internal bonding can be gained from the dissociation behaviour of the proteins. For example, all three proteins can be dissociated with detergents or urea into their 6 sub-units, and for one at least an intermediate stage, the half molecule, can be produced by relatively mild conditions [1]. In all cases further dissociation involves the disruption of disulphide bonds between the acidic and basic component polypeptides of the sub-units. Thus, qualitatively, they must all have a similar mixture of hydrogen, hydrophobic and covalent bonding. More quantitative information can be obtained from studies of their thermal denaturation using the technique of differential scanning calorimetry (d.s.c.).

The d.s.c. technique makes use of the fact that denaturation of the protein involves the absorption of energy. This energy uptake is monitored calorimetrically as the protein is subjected to a programmed rate of heating. This then yields the characteristic temperature of denaturation as well as the enthalpy of denaturation. Thus for soya 11S globulin, under specific conditions, a single transition is observed with a denaturation temperature of 94 °C (Figure 5). The latter represents the temperature of maximum heat input (T max). Also quoted sometimes is the extrapolated onset temperature (Tm) obtained by extrapolation of the initial slope of the transition to the baseline (88 °C for soya 11S). The denaturation enthalpy (ΔH) as given by the area under the transition peak is 5.6 cal g^{-1}. Both these denaturation parameters reflect the energy required to disrupt the internal bonding of the protein during unfolding and therefore must be related to the number and type of bonds present. If the internal structure and bonding of the other 11S proteins are comparable to those of soya 11S, then similar transition characteristics would be expected, and, as Table 3 shows, this is indeed the case.

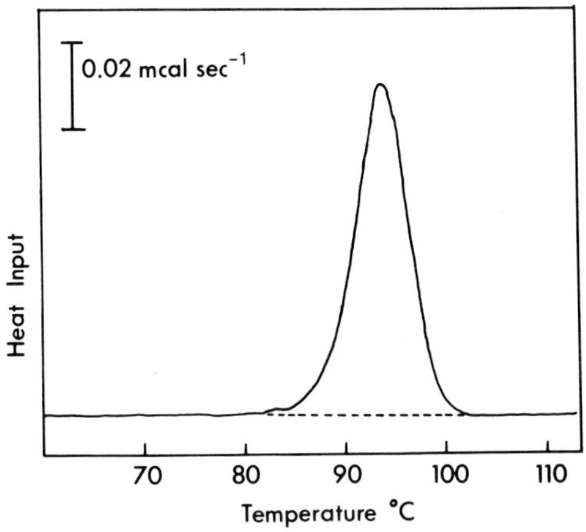

Figure 5. D.S.C. thermogram of glycinin in 0.17 *M* NaCl, 0.5 *M* NaH_2PO_4 pH 8.0 obtained by the method of Wright et al. [21].

Table 3. Thermal denaturation parameters of 11S globulins [22]

11S globulin	*T*m[a] (°C)	*T*max[a] (°C)	Δ *H*[b] ($cal\,g^{-1}$)
Glycinin (*G. max*)	88	94	5.6
Legumin (*P sativum*)	85	97	5.4
Legumin (*V. faba*)	85	94	5.4

[a] 0.17 *M* NaCl, 0.05 *M* NaH_2PO_4 (pH 8.0).
[b] 0.5 *M* NaCl, 0.05 *M* NaH_2PO_4 (pH 8.0).

Functional properties

All the physico-chemical properties discussed appear to mirror directly the similarities displayed by the proteins in their structural properties and the obvious assumption would be that this correspondence would extend to the functional properties of the three proteins. To test the validity of this argument, two functional properties of these proteins have been measured and compared. The first is the solubility, or more correctly dispersibility, behaviour as a function of pH and the second is the emulsifying capacity. Each can be considered as reflecting different aspects of the physico-chemical properties discussed. For example, the solubility or dispersibility properties will be determined principally by the size, surface charge and hydrophobicity of the proteins whilst for emulsifying capacity, hydrophobicity will be important, as well as, perhaps, the structural stability of the protein, as emulsification is thought to involve denaturation of the protein at the oil/-water interface.

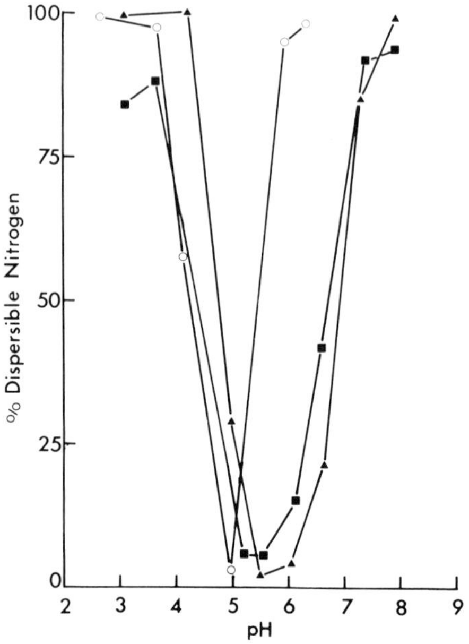

Figure 6. Dispersibility curves of the 11S globulins [23] determined by the method of Fan and Sosulski [24] with slight modifications. ○ ○: glycinin; ■ ■: legumin (pea); ▲–▲: legumin (field bean).

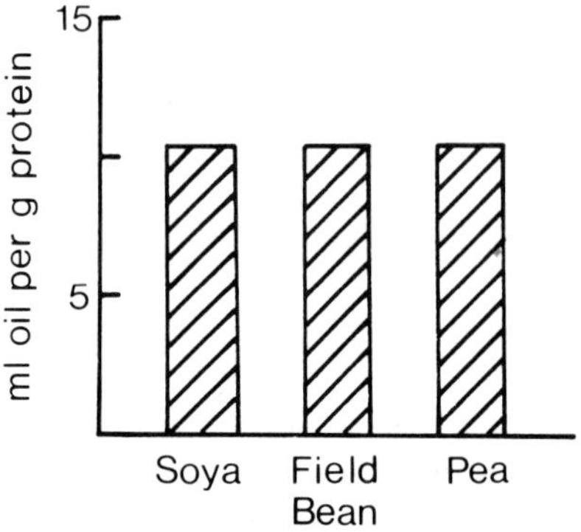

Figure 7. Emulsifying capacity of the 11S globulins [23] at pH 7.5 obtained by a modification of the method of Inklaar and Fortuin [25].

The dispersibility profiles of the three proteins in water are shown in Figure 6. All curves are of a similar shape with minima at or around pH 5, reflecting the precipitability properties of these proteins near their isoelectric points. In addition, however, each of the curves for pea and field bean 11S proteins are shifted to a higher pH compared to that of the soya 11S protein. This may be a reflection of the higher isoelectric points of the former, or perhaps of some other physico-chemical factors which have not been considered, e.g. surface reactive groups and aggregation behaviour.

As for the emulsifying properties of these proteins, Figure 7 illustrates some typical results obtained in experiments conducted with sunflower oil. All three proteins are seen to have the capacity to emulsify approximately equivalent amounts of oil.

Summary

Two conclusions can be drawn from the data presented here. Firstly, the initial assumption that there exists a relationship between structural, physico-chemical and functional properties of these proteins appears to have been borne out in general, although more extensive studies are obviously required to provide complete substantiation. Secondly, substitution of one vegetable protein for another would appear to present few problems because of the broad similarity apparent in the functional properties of these proteins that have been studied so far. These conclusions need to be qualified however, by a number of comments. Firstly, although the intrinsic properties of these proteins are seen to be roughly equivalent, other factors may affect the expression of such properties in the raw material. For example, differences in the properties of the 7S and 11S storage proteins will inevitably mean that the ratio of 7S and 11S protein will play a significant role in determining the overall properties of the material. Variation in the relative content of these two proteins in the three legumes has been reported [26–28]. In addition, other non-protein components present in the seeds (e.g. lipid) may interact with the proteins and thereby alter their basic properties. Furthermore, as discussed elsewhere, the conditions used for the extraction and preliminary processing of these proteins can have a considerable influence on their subsequent behaviour.

Finally, the structure/function relationship that I have attempted to demonstrate here is based purely on *qualitative* comparisons. Unless some *quantitative* relationships are established between functional and physico-chemical or structural properties, then it will not be possible to predict with confidence the functional behaviour of any 'protein X'. The extensive homology apparent in the structures of these vegetable proteins may provide us with just that opportunity to study the precise relationship between the structure and function of proteins.

References

1. Derbyshire E, Wright DJ, Boulter D (1976) Review. Legumin and vicilin, storage proteins of legume seeds. Phytochemistry 15:3–24
2. Mori T, Utsumi S (1979) Purification and properties of storage proteins of broad bean. Agric Biol Chem 43:577–583

3. Casey R (1979) Immunoaffinity chromatography as a means of purifying legumin from *Pisum* (pea) seeds. Biochem J 177:509–520
4. Badley RA, Atkinson D, Hauser H, Oldani D, Green JP, Stubbs JM (1975) The structure, physical and chemical properties of the soy bean protein glycinin. Biochim Biophys Acta 412:214–228
5. Lotan R, Siegelman W, Lis H, Sharon N (1974) Subunit structure of soybean agglutinin. J. Biol Chem 249:1219–1224
6. Koshiyama I (1968) Chemical and physical properties of a 7S protein in soybean globulins. Cereal Chem 45:394–404
7. Gilroy J, Wright DJ, Boulter D (1979) Homology of basic sub-units of legumin from *Glycine max* and *Vicia faba*. Phytochemistry 18:315–316
8. Moreira MA, Hermodson MA, Larkins BA, Nielson NC (1979) Partial characterization of the acidic and basic polypeptides of glycinin. J Biol Chem 254: 9921–9926
9. Casey R, March JF, Francis R, Sanger E (1979) The partial N-terminal amino acid sequence of the β-sub-units of legumin from *Pisum sativum*. Ann Report John Innes Inst: 42–44
10. Wright DJ, Boulter D (1974) Purification and sub-unit structure of legumin of *Vicia faba* L (Broad Bean). Biochem J 141:413–418
11. Catsimpoolas N, Kenney JA, Meyer EW, Szuhaj BF (1971) Molecular weight and amino acid composition of Glycinin sub-units. J. Sci Food Agric 22:448–450
12. Krishna TG, Croy RRD, Boulter D (1979) Heterogeneity in subunit composition of the legumin of *Pisum sativum*. Phytochemistry 18:1879–1880
13. Fukushima D (1968) Internal structure of 7S and 11S globulin molecules in soybean proteins. Cereal Chem 45:203–224
14. Koshiyama I (1972) Purification and physico-chemical properties of 11S globulin in soybean seeds. Int J Peptide Protein Res 4:167 176
15. Croy RRD, Derbyshire E, Krishna TG, Boulter D (1979) Legumin of *Pisum sativum* and *Vicia faba*. New Phytol 83:29–35
16. Kitamura K, Takagi T, Shibasaki K (1976) Subunit structure of soybean 11S globulin. Agric Biol Chem 40:1837–1844
17. Blagrove RJ, Lilley GG, Davey R (1980) Molecular weight of legumin from *Pisum sativum*. Aust J Plant Physiol 7:221–225
18. Danielsson C-E (1950) An electrophoretic investigation of vicilin and legumin from seeds of peas. Acta Chem Scand 4:762–771
19. Shutov AD, Vaintraub IA (1966) Chromatographic isolation and some properties of the legumin and vicilin of vetch. Biokhimiya 31:726–735
20. Bigelow CC (1967) On the average hydrophobicity of proteins and the relation between it and protein structure. J Theor Biol 16:187–211
21. Wright DJ, Leach IB, Wilding P (1977) Differential scanning calorimetric studies of muscle and its constituent proteins. J Sci Food Agric 28:557–564
22. Wright DJ, Bacon JR Unpublished results
23. Wright DJ, Bumstead MR Unpublished results
24. Fan TY, Sosulski FW (1974) Dispersibility and isolation of proteins from legume flours. Can Inst Food Sci Technol J 7:256–259
25. Inklaar PA, Fortuin J (1969) Determining the emulsifying and emulsion stabilizing capacity of protein meat additives. Food Technol 23:103–107
26. Wolf WJ, Babcock GE, Smith AK (1961) Ultracentrifugal differences in soybean protein composition. Nature 191:1395–1396
27. Gatehouse JA, Croy RRD, McIntosh R, Paul C, Boulter D (1980) Quantitative and qualitative variation in the storage proteins of material from the EEC Joint Field Bean Test. In: Bond DA (ed), *Vicia faba:* Feeding Value, Processing and Viruses. The Hague: Martinus Nijhoff, pp 173–190
28. Casey R, Sharman JE, Wright DJ, Bacon JR, Guldager P (1982) Quantitative variability in *Pisum* seed globulins; its assessment and significance. Qual Plant Plant Foods Hum Nutr 31:333–346

Extraction processes and their effect on protein functionality

P.J. LILLFORD

Unilever Central Resources, Colworth House, Sharnbrook, Bedford, U.K.

Key words: extraction processes, protein functionality, vegetable proteins

Abstract. The refinement of vegetable protein sources involves processes which manipulate the mechanical and physicochemical properties of the source materials to obtain selective removal or destruction of undesirable components. However, the choice of conditions of temperature, water content, pH, solvent hydrophobicity etc., does not necessarily identify the eventual functional properties of the proteins in human foods as a primary constraint. This can produce less than optimal raw materials, and uncontrolled damage to the proteins during extraction, concentration or drying, making cross comparison of the intrinsic properties of different source materials quite meaningless. Alternatively, the proper choice of the process conditions used, with respect to the physical chemistry of the component macromolecules, can produce a wide variety of protein states from the same source material, and hence a spectrum of functional performance. While the details of process control for a given protein source may not be exactly defined, the principles are understood. The major problem now concerns the specification of the properties appropriate to a given functional use.

Raw materials

The choice of processing route and the properties of the final product are at least in part determined by the raw material, i.e. its species, seasonal variations in the crop, the internal structure of the particles and any previous processing applied to it.

Species variation

Soya beans are considered a bulk commodity, and for oil extraction only the yield and oil content and not the distribution of protein types has determined the choice of species. However, it has been shown quite clearly that changing ratios of the 7S and 11S storage proteins exist between species. After the same processing (e.g. tofu formation), the product properties are quite different [15].

The economic dominance of the oil fraction does not apply to peas and beans. Variation in the commercially grown species is reflected by variations in the distribution of vicilin and legumin (7S and 11S) fractions of the storage proteins, so that much greater potential flexibility in the final protein enriched products exists [24].

The data on rapeseed storage proteins are less well surveyed. Both *B. campestris* and *B. napus* are grown commercially and both contain at least two major proteins (1.7 and 125 globulins) but variation in their distribution is not used or considered in chosing species for growth. Oil yield and toxicity are still the major factors.

Seasonal variation

If vegetable protein is to be recovered from seeds, the production by nature of the mature seed provides a convenient control process for arresting biological change. Examination of the developing seed [4] or of its subsequent germination [1] show significant changes in the protein distribution, but as yet neither of these processes have been used to control food material properties.

In the case of vegetable protein derived from leaves or other 'growing' parts of the plant, greater intrinsic variation between new and old growth are significant. However, yield and nutritional value have determined the cropping, rather than the potential functional properties of the combination of proteins present [12].

Preprocessing

The bulk of oil seeds are still primarily processed for their oil. For the larger seeds (soya, groundnut), hexane extraction involving relatively low temperature (60–80 °C) is used. Neither these temperatures, nor the use of the non-polar solvent cause denaturation of the protein so that defatted meals of high NSI (nitrogen solubility) are readily available.

Rapeseed, because of its size, is regarded as one of the most difficult sources to oil extract. Most current processes maximise extraction by compressing the residual cake at temperatures greater than 100 °C. Such temperatures do not necessarily denature the protein [9]. The denaturation temperatures can be elevated by as much as 40 °C if the protein is dry rather than hydrated [14]. However, one of the most economic methods of providing heat is still via 'live' steam, which also increases humidity and therefore the chance of denaturation.

Refining processes

An important constraint on refining is its cost. Only if the value added to the refined product is sufficiently high is the refinement process commercially feasible. Refining processes are capital and energy intensive, benefits therefore derive from economy of scale. These further constraints favour simple processes with few process steps, low energy consumption and a stable supply of raw material in large tonnages. The latter constraint is readily fulfilled by vegetable protein, but what processes are required or possible? These can be summarized as:

Mandatory: removal of antinutritional factors, e.g. Trypsin inhibitors, phytohaemagglutinin, goitrogen, saponins.

Desirable: removal of indigestible sugars, i.e. the flatulence factors raffinose, stachyose, verbascose.

Optional: protein isolation, fractionation, specific modification.

These requirements are obviously not mutually exclusive so that a route to an isolate may intrinsically contain steps which fulfil other requirements. Again the intrinsic properties of the raw material influence the economics of the eventual process chosen.

The refinement processes currently in operation can be further classified as non-aqueous or aqueous. Whilst providing admirable solubility properties for the separation of components, the major disadvantage of water is that since most refined products are transported and sold dry, every drop added must be removed, at a cost. Mechanical separation of dry materials or the use of more volatile solvents is obviously desirable.

Mechanical separation

Air classification. The possibility of producing carbohydrate and protein enriched fractions by direct air classification of milled legume flours has been ably demonstrated [23, 18]. The constraints on the process appear to be (i) low lipid content in the flours, otherwise agglomeration occurs, (ii) carbohydrate in the form of starch granules – otherwise classification fails. The process does not modify the native state of either the carbohydrates or proteins. However, provided the materials are used in products where heating will destroy the labile inhibitors and phytohaemagglutinins, the hazards to health or growth are no higher with these refined products than with the whole seed or meal. Furthermore, should fractions be required for which nutritional values are higher, a combination of air classification together with dry roasting processes such as those already described for Navy bean [25] might prove interesting. Further work is also worth considering to examine whether alternative milling procedures could produce classifiable fractions from defatted soya, groundnut, lupin, etc. [19].

Liquid cyclone. A process for 'sorting' the gossypol containing pigment glands of cottonseed from its finely ground meal has been developed. The process essentially uses the differential sedimentation of the pigment glands and protein bodies in hexane [22]. The objective was to achieve the mandatory detoxification of glanded cottonseed. The result, however, is a very high protein flour, almost classifiable as a concentrate (66% protein). The cyclone process should not alter the protein native state. However, the subsequent desolventisation, drying and sterilisation can damage the protein if conditions are not maintained to prevent thermal denaturation. In view of the significantly different morphology of the glands and the protein bodies, it is possible that a simpler air classification process would achieve the same results, with less chance of damaging the storage proteins.

Use of mixed solvents

The use of alcohol/water mixtures is well known for concentrate production. Of the protein components, only the Bowman–Birk inhibitor has significant solubility, while the lower molecular weight sugars, flavour precursor and saponins are removed. Provided an appropriate alcohol/water mixture is employed, the extraction process need not denature the protein. However the thermal stability of the storage proteins are known to be reduced in such media [3]. Desolventisation, therefore, frequently denatures the proteins if cost criteria define the process.

Refinement by aqueous extraction

An enormous variety of refining processes have been developed, producing concentrates and isolates from a wide variety of source materials. All depend on the differential solubility of protein, sugar, and undesirable or toxic small molecules in water, as a function of pH, temperature or ionic composition of the media. Specifics of these processes cannot be discussed here. Instead the properties of the products will be related to the physical chemistry of the various combinations of process steps. An overall scheme is shown in Figure 1.

Several research careers in protein physical chemistry have, and will be, spent identifying factors controlling the various process steps outlined above.

Concentrates. Both processes require the insolubility of protein so that the unwanted sugars etc. can be aqueous extracted. Since low protein losses are required, acid pH in the isoelectric region of the dominant proteins are chosen for washing (pH 4–6) and temperatures to thermally denature all of the proteins ($> 90\,^{\circ}C$) are used. Subtle choices of the wash pH, denaturation temperature, and pH at which it is carried out, will all yield concentrates of slightly different functional properties. But all of these are dominated by the structure and composition which is under process as well as by biological control. For example, acid leached concentrates will contain soluble protein provided the post-wash drying is mild. Their surfactant activity will therefore be higher than heat processed material but their water retention lower unless used in a subsequent process in which their heat gelation capabilities are employed.

Isolates. The pH dependence of protein extractability and precipitation is known for many vegetable proteins (e.g. [2]). During extraction at extremes of pH ($2 > pH > 11$) the extracted proteins are subject to denaturation, so that even though isolates are obtained, their subsequent functional properties will be markedly different, showing lower solubility, and less heat gelation power than the native protein types.

Many of the proteins contain labile disulphide bonds, so that the incorporation of a reducing agent in the extract medium can produce higher

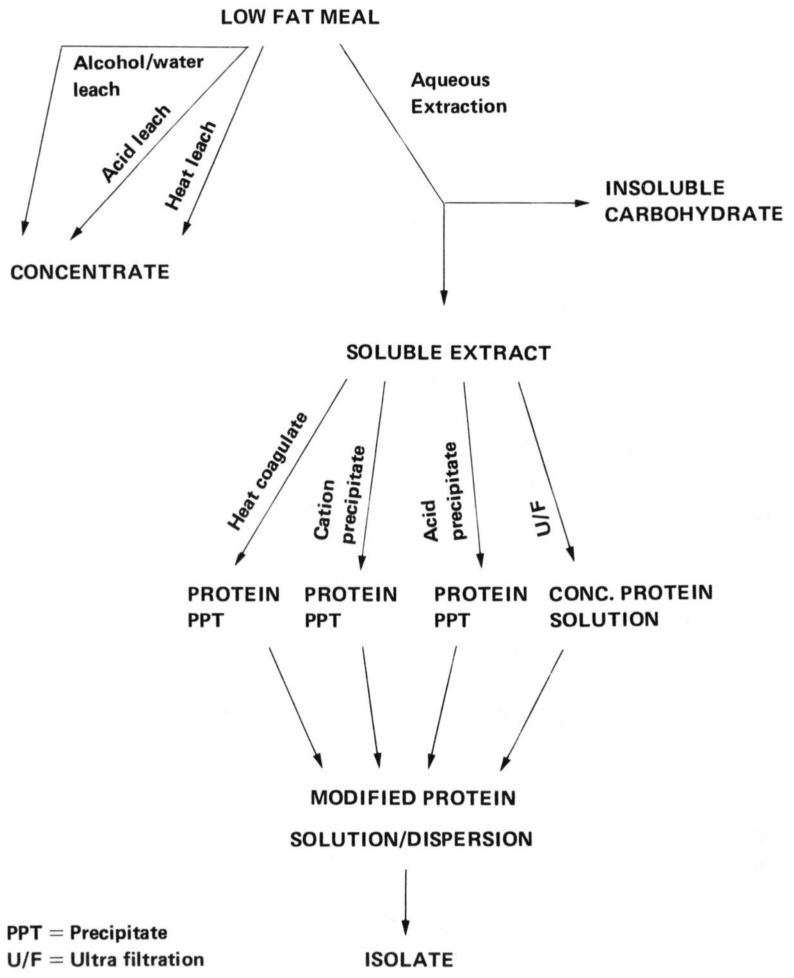

Figure 1.

yields and a subsequently more soluble and less aggregated isolate (e.g. [21, 16]). The latter process [21, 16] uses both reducing agents and the differential solubility properties of the 7S and 11S globulins to produce a particular functional performance.

Large scale 'salting out' methods for protein recovery are not practised commercially since the by-product recovery costs are too high. Precipitation by specific cations (e.g. Ca^{++}) is practised and forms the basis of Japanese tofu manufacture. The precipitate formed with soya is rubbery and not easily resolubilised, therefore making this route undesirable for isolates. Heat coagulation is also possible. The isolate is necessarily denatured and its subsequent functional use is therefore limited. Where nutrition or digestibility is the required property, however, such materials are quite appropriate, so that heat coagulation is widely practiced in leaf protein recovery [13].

Isoelectric precipitation is by far the most commonly used recovery method. The study of soya has been exhaustive, but comparable data for yields of precipitate and whey proteins from other sources are available (e.g. [17]). Precipitation and the subsequent centrifugal concentration of the proteins can lead to extensive aggregation and modification of the solubility properties. This is not solely due to disulphide interchange but also to specific protein/protein electrostatic interactions [6]. Careful control of the precipitation pH allows protein fractionation to be performed so that functional characteristics of separated proteins can be exploited [20].

These deleterious effects can also be avoided by the use of ultrafiltration rather than precipitation as a recovery method. The solubilised proteins can be concentrated to levels as high as 25% which makes direct drying of the retentate commercially competitive. Processes for the production of isolates from soy [10], sunflower and rapeseed [7], cottonseed [5] and faba bean [11] have all been described. Fractionation can be achieved if selective membranes are available. The protein can be maintained in an essentially native state throughout the process, allowing the greatest freedom in optional modification for subsequent fractional use. Outline costings suggest that such processes are quite competitive with the precipitation process [11]. Concentrates containing native protein can also be produced by a similar route [5].

All of the above comments on potential differences in the state of the proteins may be overruled by excessive drying temperatures. In general, the 7S globulins of legumes and oil seeds can be denatured at temperatures as low as 70° [24]. Therefore hot air drying on drums or fluidised beds extensively denatures the proteins. The dried particles are effectively small pieces of dried gel which may therefore have significant cold gelling, or water retention properties, largely controlled by the pH at which the drying is performed, but very low solubility. Spray drying appears to be the most reliable method of producing undamaged isolate. The temperature of the drying droplet does not increase until the water content drops to values at which the denaturation temperature is elevated. Even so, the high concentrations produced during drying generally lead to irreversible aggregation and loss of true solubility.

Ingredient/product compatibility

From the foregoing discussion it is apparent that the refined protein products potentially available to us are enormous both in number and in potential behaviour. From the process developer's point of view this can be seen either as a curse of process and raw material control or as an opportunity to produce a myriad of refined ingredients, all with different potential properties. The value of an ingredient, however, is not judged with respect to its novelty but rather its ability to fulfil a market demand. Unfortunately, a cursory examination of the physical chemistry of more traditional sources (e.g. meat, fish, egg, milk) shows such a diversity, that we can predict that refined materials exhibiting a spectrum of properties is required. But what is the spectrum and which part is relevant to which application?

Suppose we write the 'functionality' of a traditional protein in a single product as a combinational function, e.g.:

$F = A$ [solubility] $+ B$ [solution viscosity] $+ C$ [thickening] $+ D$ [heat setting] $+ D$ [surface stability] $+ F$ [rate of adsorption] ...etc.

For any product we *do not* know (1) the parameter list, or (2) the weighting factors. But we *do* know that the parameters themselves are each a function of temperature, pH, mol. wt., etc., i.e. process variables. Since ingredient users are less likely to change their process than their ingredients, we now need to identify the ingredient with the material it is intended to replace and through functional tests attempt to compare the parameters of the traditional and novel materials. Even when such measurements are carefully performed, the success of the replacement ingredient will depend on (1) our best guesses as to what the original ingredient was doing – and hence the test methods applied, and (2) the weighting of these properties in any particular product – which necessitates in-product testing.

Conclusions

The approach developed by Ralston Purina in identifying specific isolates for special uses [8] demonstrates that considered processing can produce an enormous spectrum of end-use from a single raw material. If we develop the approach further, there is no need to limit ourselves to one single replacement ingredient. If the properties required cannot be obtained from a single novel ingredient, then by combining properties of two or more ingredients with the appropriate weights, a better fit to the 'functionality equation' might be obtained.

I believe that the evidence already available suggests that the useful range of novel materials is not limited primarily by differences in raw material sources nor by our ability to refine and manipulate them, but rather our ability to understand their end use. This is true with perhaps one exception: occasionally, the function of a traditional ingredient may exploit an intrinsic feature of its protein structure, e.g. thermal gelation of egg white, anisotropy of meat and fish structures. However, the latter problem has been tackled by the production of textured ingredients (extrudates and spun fibres).

The former is a consequence of the primary structure of the proteins involved. Do we look for raw materials whose intrinsic properties are similar or 'engineer' ingredients which exhibit the same properties?

Acknowledgement. The author gratefully acknowledges the assistance of Mr CRT Brown in the preparation of this paper.

References

1. Catsimpoolas N, Ekenstam C, Rodgers DA and Meyer EW (1968) Protein subunits in dormant and germinating soybean seeds. Biochem Biophys Acta 168:122–131
2. Fan TY and Sosulski FW (1974) Dispersibility and isolation of proteins from legume flours. Can Inst Food Sci Technol J 7(4):256–259
3. Fukushima D (1969) Denaturation of soybean proteins by organic solvents, Cereal Chem 46:156–163
4. Hill JE and Breidenbach RW (1974) Accumulation of the major protein components during seed development and maturation. Plant Physiol, 53:747–751
5. Lawhon JT, Mulsow D, Cater CM and Mattil KF (1977) Production of protein isolates and concentrates from oilseed flour extracts using industrial ultrafiltration and reverse osmosis systems. J Food Sci 42(2):389–394
6. Lillford PJ and Wright DJ (1981) Influence of isoelectric precipitation on the solubility of soya bean proteins. J Sci Food Agric 32:315–327
7. Maubois J-LJ (1974) German patent 2443570
8. McAuley B (1975) The widening range of available soy protein isolates. Food Trade Review June, 15–18
9. Mustakas GC, Albrecht WJ, Bookwalter GN, McGhee JE, Kwolek WF and Griffin EL Jr (1970) Food technology, 24:102–108
10. Okubo K, Waldrop AB, Iacobucci GA and Myers DV (1974) Preparation of low-phytate soybean protein isolate and concentrate by ultrafiltration. Cereal Chem, 52(3):263–271
11. Olsen HS (1977) Continuous pilot plant production of bean protein by extraction, centrifugation, ultrafiltration and spray drying, 4th International Congress in Scandinavia on Chemical Engineering, File A5156 51–78
12. Ostrowski-Meissner HT (1980) Quantities and quality of protein extracted from pasture herbage using heat precipitation or ultrafiltration procedures (1980) J Sci Food Agric 31:177–187
13. Pirie NW (1976) Leaf protein. In: Birch (ed), Food from Waste. Applied Science, London
14. Ruegg M, Moor U and Blanc B (1975) Hydration and thermal denaturation of β-lactoglobulin, a calorimetric study. Biochim Biophys Acta 400:334–342
15. Saio R, Kamiya M and Watenabe T (1969) Food processing characteristics of soybean 11S and 7S protein. Agr Biol Chem 33:1301–1308
16. Shemer M, Creinin HL, McDonald RE and Irwin WE (1978) Functional properties of a new soy isolate. Cereal Chem, 55(3):383–391
17. Sosulski FW and Bakal A (1969) Isolated proteins from rapeseed, flax and sunflower meals, CIFT Journale 2:28–32
18. Sosulski FW and Youngs CG (1979) Yields and functional properties of air classified protein and starch fractions from eight legume flours. J Am Oil Chem Soc, 57:292–295
19. Swain RB (1975) US Patent, 3 895 003
20 Thanh V and Shibasaki K (1976) Major proteins of soybean seeds. J Ag Food Chem, 24(6):1117–1121
21. Tombs MP (1972) British Patent 1 265 661
22. Vix, HLE, Eaves PH, Gardner HK Jr and Lambou MB (1971) Degosspolised cotton seed flour – The liquid cyclone process. J Am Oil Chem Soc, 48:611–615

23. Vose JR, Basterrechia MJ, Govin PAJ, Finlayson AJ and Youngs GG (1976) Cereal Chem, 53:928
24. Wright DJ and Boulter D (1981) Differential scanning calorimetry of meals and constituents of some food grain legumes. J Sci Fd Agric, 31:1231–1241
25. Yadav NR and Liener IE (1977) Optimising the nutritive value of the proteins of navy bean by complementation with cereal proteins. Legume Research, 9(1): 17–26

Improvement of the functionality of vegetable proteins by controlled enzymatic hydrolysis

JENS ADLER-NISSEN, SVEND ERIKSEN and HANS SEJR OLSEN

Novo Industri A/S, Enzyme Application R & D, DK-2880 Bagsvaerd, Denmark

Key words: enzyme, food protein, proteolysis, functionality, pH-stat

Abstract. Vegetable proteins are available to the food industry in various forms, such as flours, concentrates, isolates, and TVP (textured vegetable protein). However, the functional properties of these products are not always optimal and need improvement for certain applications. In recent years it has been demonstrated that a limited enzymatic hydrolysis offers a convenient and specific means of improving certain functional properties of vegetable proteins. The hydrolysis reaction is controlled by the degree of hydrolysis (DH) which is defined as the percentage of peptide bonds cleaved. Under neutral or slightly alkaline conditions, DH can be monitored continuously by the pH-stat technique, since the amount of base necessary to maintain a constant pH is proportional to DH. The reaction is terminated at a preset DH-value to obtain the desired properties of the hydrolysate.

Protein products from soya bean, faba bean, and potato were hydrolysed to DH 0%, 3% and 5%, and certain functional properties (iso-electric solubility, emulsification capacity, whipping expansion, and foam stability) were evaluated. The general picture observed is an increase, often pronounced, in these four properties with increasing DH – however, there are also distinct differences between the protein sources as well as between the different forms of a particular protein. This work demonstrates that protein sources indigenous to Europe are fully comparable to soya protein with respect to potential functional uses.

Introduction

The use of refined or functional vegetable proteins in the European food industry has been increasing during recent decades. This increase is mainly due to the economical advantage of substituting expensive animal protein with low-cost vegetable protein. Technological difficulties in maintaining established assessments of good quality are encountered, however, and the applications of vegetable proteins are still limited to rather traditional products. Hence, a more widespread use of these proteins has to await changes in consumer adaption and habits and are not governed by economics alone.

It is not the intention in the present work to expound our views on the above development in general. Rather, we would like to point out that among all the various problems which might be discussed, we must not forget that the functionality conveyed by the existing vegetable protein products cannot optimally fulfil all existing needs.

Functional vegetable proteins are sold as powders, or are further processed into texturized products. The present work only deals with the possibilities of improving some functional properties of the powdered products. Apart from wheat gluten, these are today virtually exclusively derived from soybeans. Other vegetable sources of functional proteins are potentially available, but when processed into traditional concentrates and isolates they generally do not reveal functional properties that are superior to those of soybean proteins.

One of the ways of expanding the range of functional properties offered by protein flours, concentrates and isolates is enzymatic modification. For years enzymatic modification has, however, only been used commercially to a limited extent, for example in the production of whipping agents from soy isolate [7]. This hesitation in utilizing an otherwise obvious technology is mainly due to bad experiences with the well-known formation of bitter off-flavour in protein hydrolyzates.

The bitter taste of protein hydrolyzates has been a subject of study for years, and it would be outside the scope of the present work to attempt any form of review of this field here. It is generally acknowledged that the bitterness is caused by peptides with a high content of hydrophobic amino acids and usually of low molecular weight [6]. The bitterness evolves during the course of the proteolytic reaction and its intensity seems to depend mainly on the amino acid composition and on the extent of proteolytic degradation with other factors such as the choice of enzyme and presence of ions playing a smaller role [3].

The concept of controlled enzymatic hydrolysis

In recent years it has been demonstrated that if the extent of hydrolysis is limited, it is possible to achieve an improvement in functionality [3, 4, 5, 13, 17]. By limiting the extent of hydrolysis it is also possible to avoid the bitterness which becomes a problem [3]. This approach calls for a careful control of the proteolytic reaction, because the natural tendency will be that the reaction proceeds beyond the optimal stage to give bitter tasting hydrolyzates with an uninteresting functionality.

The parameter used in our work for controlling the hydrolytic reaction is the Degree of Hydrolysis (DH), which is defined as the percentage of peptide bonds cleaved [1]. The reaction is then terminated by inactivating the enzyme when the proper DH-value is reached. This may appear obvious, but it should be realized that in most of the work published on protein hydrolysis processes, time rather than DH is used as the controlling parameter. The advantage of using DH is that the properties of the hydrolyzate are largely determined by DH alone, independent (to a certain extent) of the choice of other parameters such as substrate concentration, enzyme

1. Opening of the peptide bond:

$$\text{-CHR'-CO-NH-CHR''-} + H_2O \xrightarrow{\text{Enzyme}} \text{-CHR'-COOH} + NH_2\text{-CHR''}$$

2. Proton exchange:

$$\text{-CHR'-COOH} + NH_2\text{-CHR''} \rightarrow \text{-CHR'-COO}^- + \overset{+}{N}H_3\text{-CHR''}$$

3. Titration of amino group:

$$\overset{+}{N}H_3\text{-CHR''} + OH^- \rightleftarrows NH_2\text{-CHR''} + H_2O$$

$pK_a \approx 7.7\ (25^oC);\ 7.1\ (50^oC)$

Ionization enthalpy: $\Delta H^o \sim 45$ kJ/mole

G-0304

Figure 1. Function of the pH stat

dosage and temperature [3, 4]. Another advantage is that when the hydrolysis is carried out at pH 7 or above, DH can be conveniently measured by the pH-stat technique: the amount of base, which is added to the reaction mixture to keep the pH constant, is proportional to DH [1]. The chemical principle of the pH-stat technique is shown in Figure 1.

The purpose of the present work is to demonstrate that a limited, controlled hydrolysis is a generally applicable procedure for altering the functional properties of vegetable proteins in a desirable way, in particular if these proteins have been denaturated during their processing to concentrates and isolates, or if the proteins in their native state lack sufficient functionality.

Materials and methods

Protein raw materials

- Acid precipitated soy isolate: Purina 500E, commercially available from Ralston Purina Company.
- Soy concentrate: Danpro H, commercially available from Århus Oliefabrik A/S.
- Acid precipitated faba bean isolate: faba isolate manufactured in pilot plant scale by acid precipitation of an aqueous extract of faba bean flour [14].
- Ultrafiltered soy isolate: Soy isolate manufactured in pilot plant scale by ultrafiltration of an aqueous extract of soy white flakes [14].
- Ultrafiltered faba bean isolate: faba isolate manufactured in pilot plant scale by ultrafiltration of an aqeuous extract of faba bean flour [14].
- Potato protein concentrate manufactured by air classification of dried potato solids [8].

Table 1. The calibration factor, $1/\alpha$, for the pH-stat[a]

pH	T °C	25 °C	30 °C	40 °C	50 °C	60 °C
	pK	7.7	7.6	7.3	7.1	6.9
6.5					5.00	3.50
7.0			5.00	3.00	2.27	1.79
7.5		2.59	2.27	1.63	1.40	1.25
8.0		1.50	1.40	1.20	1.13	1.08
8.5		1.16	1.13	1.06	1.04	1.03
9.0		1.05	1.04	1.02	1.01	1.01
9.5		1.02	1.01	1.01	1.00	1.00
10.0		1.00	1.00	1.00	1.00	1.00
10.5		1.00	1.00	1.00	1.00	1.00
11.0		1.00	1.00	1.00	1.00	1.00

[a] $\alpha = \dfrac{10^{\mathrm{pH-pK}}}{1 + 10^{\mathrm{pH-pK}}}$ and $\dfrac{1}{\alpha} = 1 + 10^{\mathrm{pK-pH}}$.

The protein in the first three raw materials are (partially) denatured, whereas the protein in the last three materials have retained their native states.

Calculation of hydrolysis curves

The hydrolysis curve for a given protein-enzyme system is obtained by plotting DH as a function of time, usually under the following standard conditions [4, 11]:

Substrate concentration, $S = 8\%$ (N x 6.25)
Enzyme dosage, $E/S = 2\%$ Alcalase 0.6 L (12) corresponding to 12 AU (Anson Units) per kg substrate protein
pH (pH-stat) = 8.0
Temperature (T) = 50 °C.

DH is calculated from the base consumption during the hydrolysis:

$$\mathrm{DH} = B \times N_b \times \frac{1}{\alpha} \times \frac{1}{MP} \times \frac{1}{h_{tot}} \times 100\%$$

where:

B = base consumption in litres
N_b = normality of the base
α = average degree of dissociation of the α-NH_2 groups (see below)
MP = mass of protein (N x 6.25) in kg
h_{tot} = total number of peptide bonds in the protein substrate (meqv/g protein) (see below).

The degree of dissociation is

$$\alpha = \frac{10^{\mathrm{pH-pK}}}{1 + 10^{\mathrm{pH-pK}}}$$

pK is the average pK-value of the α-amino groups liberated during the hydrolysis, and it can be determined from a direct assay of the liberated amino groups as described previously [4]. pK varies significantly with temperature, but is fairly independent of the substrate as such: Table 1 gives the values of $1/\alpha$ for a series of pH-values and temperatures.

h_{tot} is calculated from the amino acid composition of the protein. It is 7.8 meqv/g protein (N x 6.25) for soy protein and faba bean protein. (If the amino acid composition is not known an average value of 8 meqv/g can be assumed.) Certain raw materials, such as potato protein and single cell protein, contain considerable amounts of low molecular weight, non-protein nitrogen compounds, and for these raw materials it seems relevant to correct h_{tot} by multiplication with the TCA-solubility index [4].

In the case of the present potato protein concentrate, a TCA-solubility index of 44% was found, which means that a h_{tot} (corr.) value of 3.1 meqv/g protein (N x 6.25) is obtained in comparison to the uncorrected value of 7.1 meqv/g protein (N x 6.25).

Production of functional protein hydrolyzates

All proteins were hydrolyzed with Alcalase 0.6 L under the following conditions:

$$S = 8\% \text{ protein } (\text{N} \times 6.25) \quad E/S = 2.0\% \quad \text{pH} = 8.0 \quad \text{T} = 50\,^{\circ}\text{C}.$$

The hydrolyses were terminated at various DH values by addition of HCl to pH 4.0 to inactivate the enzyme. The pH was then readjusted to 7.0 and the product was spray-dried or freeze-dried, depending on whether it was produced in the pilot plant or in the laboratory. The final products were analysed for nitrogen (Kjeldahl), dry matter, and ash (600 °). The DH 0 products, which are used as controls, were produced without enzyme with a holding time of 15 min at pH 8.0 before lowering the pH to 4.0 and then readjusting to pH 7.0.

Evaluation of functional properties

Solubility at pH = 4.5 was determined in a 1% protein dispersion in 0.2 *M* NaCl. After stirring with a magnetic stirrer for 1 hour the suspension was centrifuged and the supernatant was analysed for nitrogen [3].

Emulsifying capacity was determined twice on each product by a modified Swift titration. The parameters used were:

Homogenizer: Sorvall Omnimixer

Procedure: 4.0 g protein (N x 6.25) in 250 ml 0.5 *M* NaCl. Blend for 2 min at low speed. Transfer 50 ml to a glass blender jar, weigh and add 50 ml soya oil. Homogenize at 10 000 rpm with jar in ice-bath. Add soya oil (0.9 ml/sed) until the emulsion collapses [4].

Whipping expansion was determined in a 3% protein solution at pH = 7. The experimental details were previously described [3].

Foam stability was calculated as the ratio of foam left (after draining for 30 minutes) to the original amount of foam [16].

Results and discussion

Hydrolysis curves

Figure 2 shows the hydrolysis curves for the six different raw materials. There is a considerable difference between the raw materials with respect to the reaction rates; in particular the ultrafiltered isolates are hydrolyzed much more slowly than the acid precipitated isolates. This difference can be ascribed to the compact structure of the native globular proteins, as discussed previously [16].

Production of functional protein hydrolyzates

To avoid using excessively long hydrolysis times the undenatured protein materials were only hydrolyzed to DH 3% in the subsequent experiments. In the case of the denatured substrates, hydrolyzates were produced both at DH 3% and DH 5%. In all cases 'blanks' were produced (without enzyme) – these are denoted DH 0.

Table 2 shows the protein and ash contents of the raw materials and the hydrolyzates produced. There is, as expected, a systematic increase in the ash content (and consequently a decrease in protein content) with increasing DH (except for the DH = 0 and 3 products in the case of ultrafiltered soy isolate where, for other reasons, salt was added to adjust the ash content to the same values). It is clearly seen from the results of the DH 0 products that a considerable part of the ash found in the DH 3 hydrolyzates originates from the inactivation procedure (pH adjustment). The ash content might thus be lowered if needed, provided the inactivation of the enzyme is carried out by a heat treatment instead of lowering the pH. For most applications the more simple acid inactivation should be adequate, as the proportion of ash to protein is still reasonably low, except in the case of potato protein concentrate.

Isoelectric solubility

Figures 3 and 4 show the isoelectric solubility of the denatured and undenatured proteins, respectively. The denatured proteins show a distinct similarity in their response to the proteolytic treatment: in all cases the solubility index is improved from around 10% to around 40% at DH 3%. Hydrolysis to DH 5% causes the solubility to increase slightly more.

It is known from previous work on soy isolate that the effect on the solubility index is most pronounced around the isoelectric point; in fact,

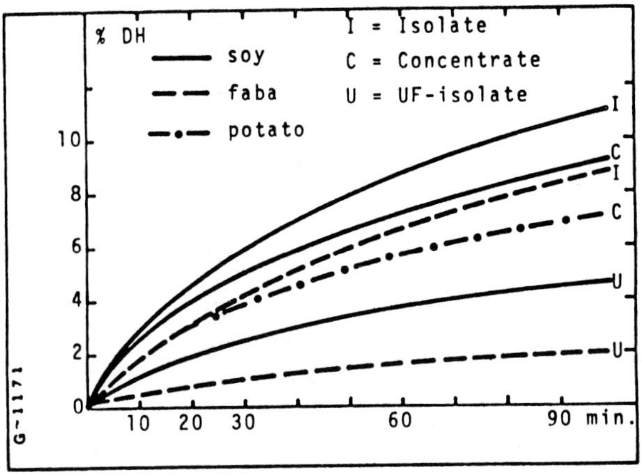

Figure 2. Hydrolysis of various vegetable proteins with Alcalase – hydrolysis curves at standard conditions.

Table 2. Dry matter composition of protein materials

Protein material	N × 6.25 %	Ash %
Acid precip. soy isolate		
Untreated	90	3
DH 0	85	8
DH 3	83	10
DH 5	82	11
Acid precip. faba isolate		
Untreated	87	7
DH 0	84	9
DH 3	79	14
DH 5	77	15
Soy concentrate		
Untreated	71	6
DH 0	67	10
DH 3	65	12
DH 5	63	12
Ultrafilt. soy isolate		
Untreated	86	7
DH 0	78	15
DH 3	76	15
Ultrafilt. faba isolate		
Untreated	87	5
DH 0	82	11
DH 3	80	14
Air classified potato protein concentrate		
Untreated	36	13
DH 0	31	17
DH 3	31	19

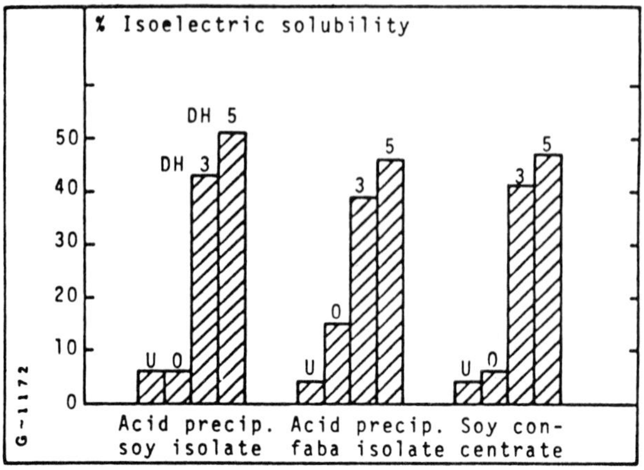

Figure 3. Denatured proteins. Isoelectric solubility.

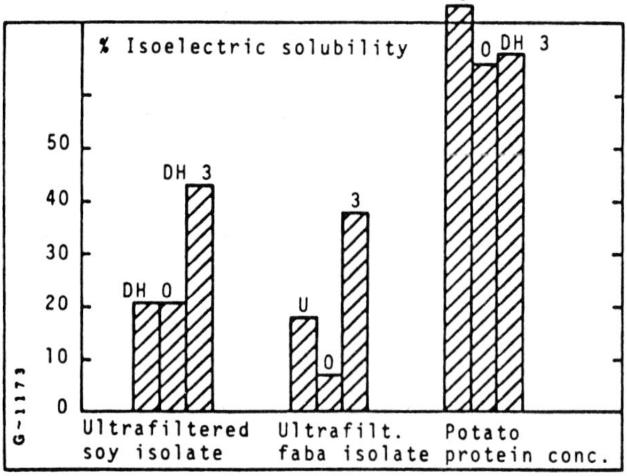

Figure 4. Undenatured proteins. Isoelectric solubility.

the proteolytic treatment changed the pH-solubility profile (solubility index versus pH) from the usual U-shape to a fairly flat plateau [3]. The flat solubility curve is of considerable advantage in many food systems as it prevents occasional precipitation problems due to otherwise unavoidable fluctuations in pH.

The solubility of the ultrafiltered isolates (Figure 4) shows a similar increase as the acid precipitated isolates. The presence of soy 'whey' proteins in the ultrafiltered soy isolate is reflected in the fairly high isoelectric solubility of 21%, compared with 6% for the acid precipitated isolate. The decline in solubility from the untreated to the DH 0 product for two of the native proteins might be explained by some denaturation during the enzyme inactivation procedure.

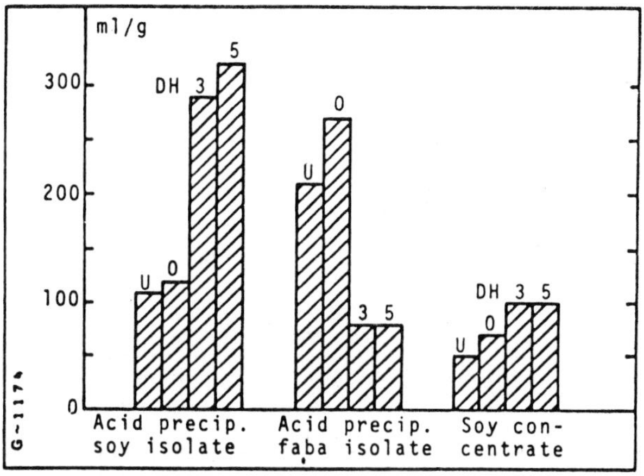

Figure 5. Denatured proteins. Emulsification capacity.

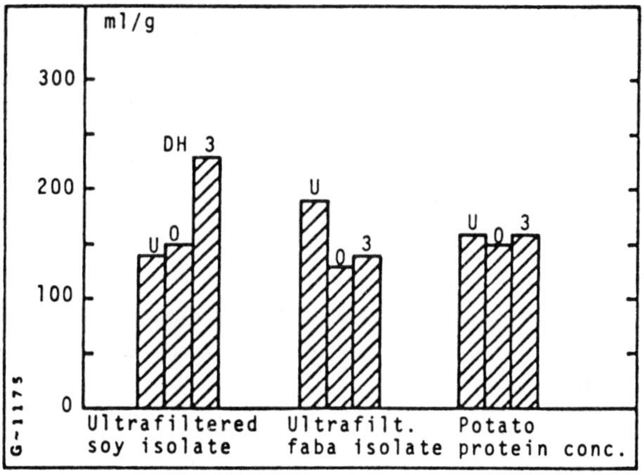

Figure 6. Undenatured proteins. Emulsification capacity.

The behaviour of protato protein concentrate is completely different from that of the other raw materials. Its isoelectric solubility is initially high and is not appreciably changed by the proteolytic treatment. The last observation may seem surprising but can be explained by substrate competition. Both the large, insoluble proteins and the smaller, soluble proteins and peptides, which are present in potato protein concentrate, serve as substrates for the enzyme. This has the consequence that a large fraction of the activity of the enzyme molecules is wasted in an attack on low molecular weight substrate which is already soluble.

Emulsification capacity.

Figures 5 and 6 show the emulsification capacity of the denatured and undenatured proteins, respectively. For soy protein products an increasing emulsification capacity with increasing degree of hydrolysis is obtained, the most striking case being the acid precipitated soy isolate. However the differences between the DH 3 and 5 products are minor. The observed emulsification capacities of acid precipitated soy isolate are quantitatively in agreement with previous measurements using the same experimental conditions [4].

For the faba protein products a decrease in emulsification capacity is noted when the proteins are hydrolyzed. This is in contrast to previous work on enzyme modified faba isolate, where a slight increase was observed [15]. A closer comparison of the results shows that the emulsification capacities of the hydrolyzed proteins are comparable in size, whereas the raw material and the control (DH 0) exhibit an unusually high emulsification capacity in the present work, both in comparison with the other proteins and with previous measurements on faba protein isolate [15]. Repeated measurements on the unhydrolyzed proteins gave the same qualitative picture albeit with considerable numerical differences. It must therefore be concluded that the influence of a proteolytic treatment on the fairly good emulsification capacity of faba protein still remains an open question.

The emulsification capacity of potato protein concentrate appears not to be affected by the proteolytic treatment. A systematic study of the emulsification properties of potato protein concentrate demonstrated the superiority of this protein in comparison with other vegetable proteins [9]. Any improvement of the emulsification capacity is therefore harder to achieve than with proteins such as acid precipitated soy isolate which has a fairly low emulsification capacity in its unmodified state.

Whipping properties

Figures 7 and 8 show the whipping properties of the denatured and undenatured proteins, respectively. The proteolytic treatment causes a dramatic increase in whipping expansion as observed in previous work [3, 4]. However, in the case of the isolates, the stability of the foam is low. The hydrolyzates of soy concentrate, on the other hand, display a remarkably high foam stability. It cannot be ruled out that the polysaccharides present in soy concentrate contribute to this stabilization. In fact, when isolated soy polysaccharides were added to the DH 5% hydrolyzate of soy isolate (1 part polysaccharide to 4 parts protein), it was found that the expansion after 30 minutes increased from 60% to 170%. This is still a low figure, however, compared with the 620% value obtained with the DH 5% hydrolyzate of soy concentrate. The apparently favourable whipping properties of hydrolyzed soy concentrate deserve more investigation since soy concentrates are attractive raw materials from an economical point of view.

The whipping properties of the undenatured protein materials are in general excellent, in particular with respect to stability. Surface denaturation is one phenomenon which can be expected to enhance foam stability and

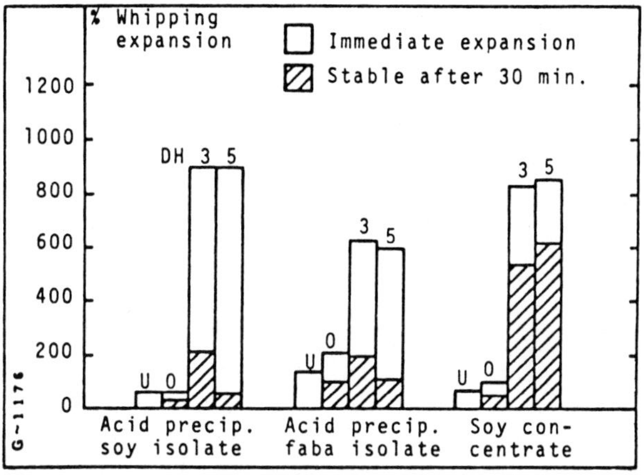

Figure 7. Denatured proteins. Whipping properties.

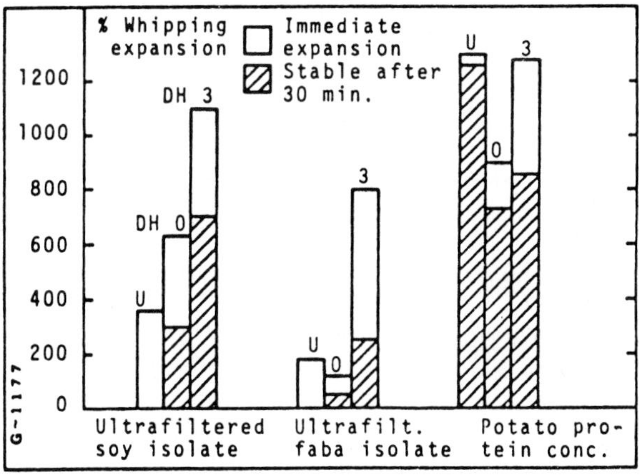

Figure 8. Undenatured proteins. Whipping properties.

this may explain why, for example, ultrafiltered soy isolate has a higher foam stability than the already partially denatured acid precipitated isolate. The relatively poor performance of faba isolate, both denatured and undenatured, may be ascribed to its relatively low content of free SH-groups [10]. For both isolates it is observed that the whipping properties are improved by a hydrolysis to DH 3%. Only the last material, potato protein concentrate, is not improved by the proteolytic treatment; in fact the contrary is observed. This illustrates that it is not possible to make quantitative predictions from one enzyme-substrate system to another with regard to the functional properties. Each combination of enzyme and substrate will have to be investigated through experimental work.

The fact that a proteolytic treatment of ultrafiltered soy isolate results in excellent foaming properties has led to further studies in this field. By proper combination of ultrafiltration and enzyme modification, whipping agents with egg-white properties have been produced [16].

Conclusions

The present work has demonstrated that a limited hydrolysis of vegetable food proteins changes their functional properties, most often in an attractive way. In particular, an improvement is seen if the protein substrate is partially denatured, as is the case with most commercially available vegetable proteins. However, there are also distinct differences between the various protein materials and the above conclusion can therefore only be stated qualitatively. It should also be stressed that the protein hydrolyzates have been investigated in model systems only, and their performance will ultimately have to be evaluated in actual food systems. However, it is to be expected that the optimal DH-value found from studies such as the present one, will also appear to be optimal in more complex food systems.

The ubiquitous soy proteins are the raw materials of choice for anybody who wants to make functional protein ingredients by controlled enzymatic hydrolysis. But we should not forget the other protein sources which are available in Europe; as the work has demonstrated it is possible to produce protein hydrolyzates with good functional properties from faba protein. Air classified potato protein concentrate offers in itself excellent whipping properties. We can therefore conclude that protein crops which are indigenous to Europe by no means should be excluded from having a future as sources of functional ingredients for the European food industry.

Acknowledgment. We wish to thank the staff at Application Technology I for their skillful practical work in connection with this publication.

References

1. Adler-Nissen J (1977) Enzymatic hydrolysis of food proteins. Process Biochem 12(6):18–23, 32
2. Adler-Nissen J (1978) Enzymatic hydrolysis of soy protein for nutritional fortification of low pH-food. Ann Nutr Alim 32:205–216
3. Adler-Nissen J, Sejr Olsen H (1979) The influence of peptide chain length on taste and functional properties of enzymatically modified soy protein. ACS Symp Ser 92:125–146
4. Adler-Nissen J (1981) Limited enzymic degradation of proteins – a new approach in the industrial application of hydrolyses. J Chem Technol Biotechnol 32:138–156
5. Bobalik JM, Taranto MV (1980) The effect of enzymic modification on the foaming, water absorption and baking quality of defatted soy flour. J Food Technol 15:637–646
6. Eriksen S, Fagerson IS (1976) Flavours of amino acids and peptides. Int Flavours Jan–Feb 1976: 13–16
7. Gunther RC (1979) Chemistry and characteristics of enzyme-modified whipping proteins. J Am Oil Chem Soc 56:345–349

8. Holm F (1980) A new system for the production of starch and protein from potato. Stärke 32:258–262
9. Holm F, Eriksen S (1980) Emulsifying properties of undenatured potato protein concentrate. J Food Technol 15:71–83
10. Horiuchi T, Fukushima D, Sugimoto H, Hattori T (1978) Studies on enzyme-modified proteins as foaming agents: effect of structure on foam stability. Food Chem. 3:35–42
11. Novo Industri A/S (1978) IB-102: Hydrolysis of food proteins in the laboratory. Bagsvaerd, Denmark
12. Novo Industri A/S (1980) B-207: Alcalase® 0.6 L. Bagsvaerd, Denmark
13. Puski G (1975) Modification of functional properties of soy proteins by proteolytic enzyme treatment. Cereal Chem 52:655–664
14. Sejr Olsen H (1978) Continuous pilot plant production of bean protein by extraction, centrifugation, ultrafiltration and spray dyring. Lebensm -Wiss u -Technol 11:57–64
15. Sejr Olsen H (1980) A survey on recent developments in industrial processing of *Vicia faba* protein. In: Bond DA (ed), *Vicia faba.* Feeding Value Processing and Viruses. Brussels and Luxembourg: ECSC, EEC, EAEC, pp 233–255
16. Sejr Olsen H, Adler-Nissen J (1981) Application of ultra- and hyperfiltration during production of enzymatically modified proteins. ACS Symp Ser 154:133–169
17. Zakaria F, McFeeters RF (1978) Improvement of the emulsification properties of soy protein by limited pepsin hydrolysis. Lebensm -Wiss u -Technol 11:42–44

Phenomena involved in the 'texturization' of vegetable proteins and various technological processes used

CLAUDE GIDDEY

BATTELLE, Geneva Research Centers 7, route de Drize, 1227 Carouge-Geneva, Switzerland

Key words: Vegetable protein, texturized protein, coagulation, denaturation, molecular orientation, texturization processes, animal protein

Abstract. Various fundamental aspects of vegetable protein texturization are discussed with particular reference to thermal coagulation, denaturation, and orientation phenomena. Some nine different classes or types of processes, which are used for texturization, are briefly described and discussed with detailed emphasis on the underlying fundamental mechanisms or phenomena involved. The texturization of mixtures containing both plant and animal protein is briefly discussed. Lastly, prospects for vegetable protein texturization in the future are considered.

Introduction

The topic to be discussed was certainly novel twenty years ago, but in my opinion it continues to be a present-day research problem. The reason is that this field is still in its infancy; in other words, its industrial and commercial development is not yet sufficiently advanced. This is due to psychological and commercial as well as to technological and qualitative factors. However, in this discussion, I would like to focus on fundamental technological aspects involved in the texturization of vegetable and combined vegetable/animal proteins.

Nutritional materials are considered to be textured if their physical characteristics differ from those of liquids. Viscous products, gels, and solid foods are textured materials. Essentially, a texture is formed by the more or less effective immobilization of the liquid medium. In this context, an example is found in the case of fat matter, the solid (or viscous) state of which results directly from its crystallization, as for ice-creams, sorbets, soft-ice and other frozen products.

In this discussion, I will focus particularly on the immobilization of the aqueous medium (in the case of non-frozen products) which constitutes 60 to 80% of re-constituted textured vegetable protein materials such as meat extenders and meat alternatives based on vegetable proteins.

Table 1. Composition of sunflower seed products[a]

		Defatted flour (%)		Concentrate (%)
Proteins (N_2 X 6,25)		55.7		74.2
Lipds		1.8		0.4
Ash		7.3		10.3
Crude Fiber		6.1		6.6
Phenols	Chlorogenic Acid	3.7	4	0.16
	Caffeic Acid	0.2		–
	Chinic Acid	0.1		–
Soluble sugars		10.1		0.2
Moisture		9.6		7.8
Non identified		5.4		0.94

[a]Kindly supplied by Assoreni (Roma).

General considerations

Basic Materials

Basic meterials, such as vegetable flours from oil seeds or leguminous plants, are not composed exclusively of proteins, but also contain carbohydrates. Whereas proteins are likely to favor texturization, carbohydrates are inert or even unfavorable. By removing the non-protein compounds either totally (protein isolates) or partially (concentrates), highly functional materials can be obtained. Table 1 shows the respective compositions of flour and concentrate from sunflower seed.

Protein structure

Let us recall that the structure of any natural protein initially depends on the sequence of the amino acids constituting its polypeptide chain or chains (primary structure). The physical (hydrogen bonds) and chemical (disulphide bridges) interactions between the different parts of the chain will determine, together with its hydrophilic and hydrophobic properties, the winding and folding that give the protein its specific conformation (secondary and tertiary structures) in this, its native undernatured state. In this form, the vegetable proteins present a globular conformation. When brought into contact with water, these proteins tend to disperse colloidally in the solvent. The dispersion can even become a solution if the pH is far enough away from the isoelectric point to permit ionization of carboxylic groups of the molecules (electrostatic repulsion effect). In native proteins, this phenomenon is reversible: the proteins can return to their initial conformation if the pH is brought back to the isoelectric point.

Retaining functionality

In the manufacture of isolates or concentrates, the method or procedure used is appropriate only on the condition that most of the protein contained in the isolate or concentrate remains in a native state, i.e. that the processes used do not substantially denature the protein. For example, concentrates obtained by a water/alcohol/leaching process generally do not meet this requirement because their partly denatured proteins cannot participate in the stabilization of artificial textures. On the other hand, concentrates obtained by combined alcohol/acid processes remain functional to a larger extent. We have obtained good results by using protein concentrates from sunflower seed prepared by this method (products kindly supplied by Assoreni, Roma).

Mechanisms

The texturization techniques include two essential steps:

1. Creation, in the hydrated medium, of a molecular or macromolecular body which is characterized by a directional structure (anisotropic).
2. Insolubilization of the components of the molecular body and formation of a sufficiently coherent reticulate system able to immobilize the hydrated phase of the product.

The texturization of vegetable proteins is based on this phenomenology. Consequently, orientation mechanisms as well as insolubilization and reticulation mechanisms of the proteins have to be studied. Firstly, I am going to examine insolubilization mechanisms, including coagulation and denaturation of the proteins. The orientation phenomenon will then be discussed in the following section (Texturization Processes).

Thermal coagulation/denaturation

Thermal coagulation may be used to stabilize artificial structures imposed on protein material. The aggregation and reticulation (cross-linking) of the molecules in the macroscopic state may be rendered insoluble and 'set' by heat treatment.

Because of the energy it supplies, heat breaks the internal interactive forces, in particular those of hydrogen bonds with water, encouraging the establishment of physical or chemical bonds between the polypeptide chains of the same molecule (intramolecular) or between different protein molecules (intermolecular). The transfer of intramolecular-SS-bonds into intermolecular -SS-bonds is a specific consequence of thermal treatment.

The thermal denaturation of proteins facilitates three modes of solidification of the protein-rich water medium:

a. Thermo-reversible gelation (gelatin type).
b. Non-reversible gelatinisation.
c. Non-reversible coagulation (aggregation).

The molar concentration of the hydrophobic amino acids of the protein chain is one of the elements which determines the type of solidification phenomenon, as has been shown by Shimada and Matsushita [6]. The molecular weight of proteins is another important factor in their behavior when heated, since initial protein monomers may form molecular groups (polymers), which can be the starting point of further coagulation.

The phenomena discussed above play a decisive part in the texturization process and in establishing the textural organoleptic characteristics of the products concerned.

Texturization processes

The characteristics of textured vegetable products depend primarily, as has been said formerly, on the orientation imparted to the constituent material, either at the molecular level, or at least, at the microscopic or submicroscopic levels. It is the primary task of the texturization process to orientate the individual protein molecules so as to confer, after setting, an anisotropic resistance to the food.

It is remarkable to observe that the numerous processes (during the past twenty years, more than 200) which have been proposed, and usually patented, include always at least one technical step aimed at imparting a discrete or important orientation to the protein material. This common aim of technologies will become apparent in the subsequent analysis of model processes.

The numerous existing processes, some of which are already industrially applied, can be divided into 9 classes which cover the majority of the proposed technologies. The differences, within the classes, are indeed more concerned with the conception of the machinery than with the principle of texturization itself. The classes, or types of processing, are:

1. Wet spinning.
2. Cooking extrusion.
3. Gel texturization.
4. Tear texturization.
5. Plastification – extrusion.
6. Melt spinning.
7. Solvent texturization.
8. Texturization by surface deformation.
9. Texturization by freezing.

Wet Spinning (1)

I will not discuss *wet spinning* in detail. To sum up, this is the only technology which involves total unfolding of the protein chains of the isolates (or of the concentrates) of vegetable proteins. The alkaline solubilization of the globular vegetable proteins (spinning solution) disposes them to fiber formation in the acid coagulating bath. The insolubilization of the components results from the acid coagulation of the protein. Thus, coagulation stabilizes the linear

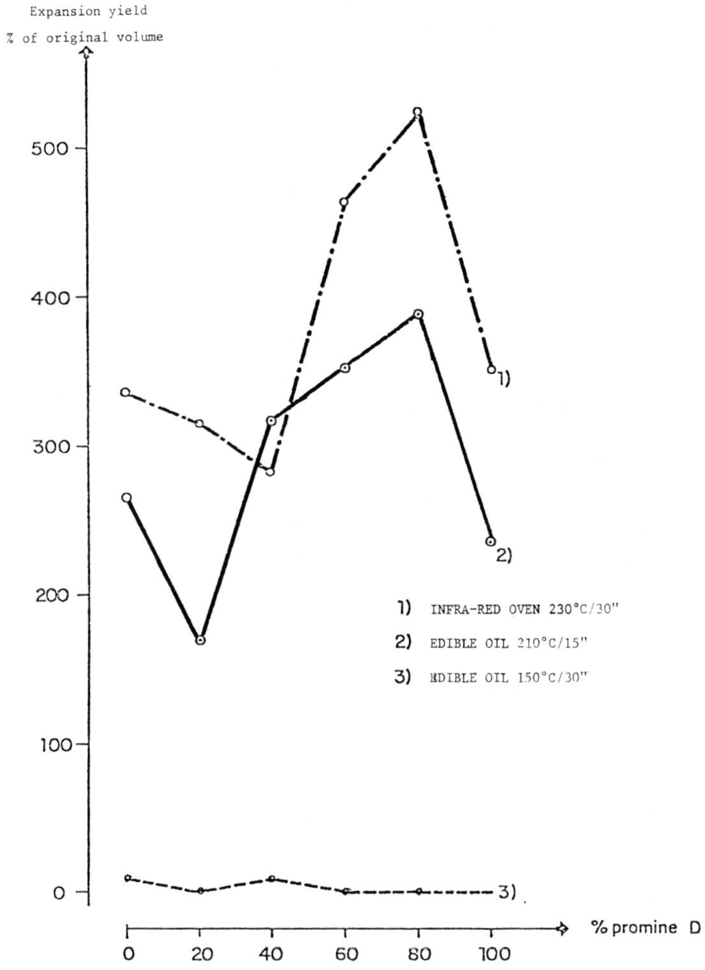

Figure 1. Expansion tests: wet intimate mix of soy protein isolate (promide D) and corn starch.

oriented structure of the macromolecules, conferring on the fibers their outstanding textural qualities.

In such a process, heat does not directly influence the coagulation itself. The coagulation instead results from neutralization of the negative charges and the reticulation promoted by the low pH which is often reinforced by a combination with reactive cations of calcium salts and/or mineral salts contained in the spinning bath.

Contrary to spinning, the other processes require no solubilization of the proteins. The proteins are merely hydrated to different extents, thereby entailing a limited alteration of their conformation. The orientation mechanisms will therefore be concerned more with the assembly of groups of

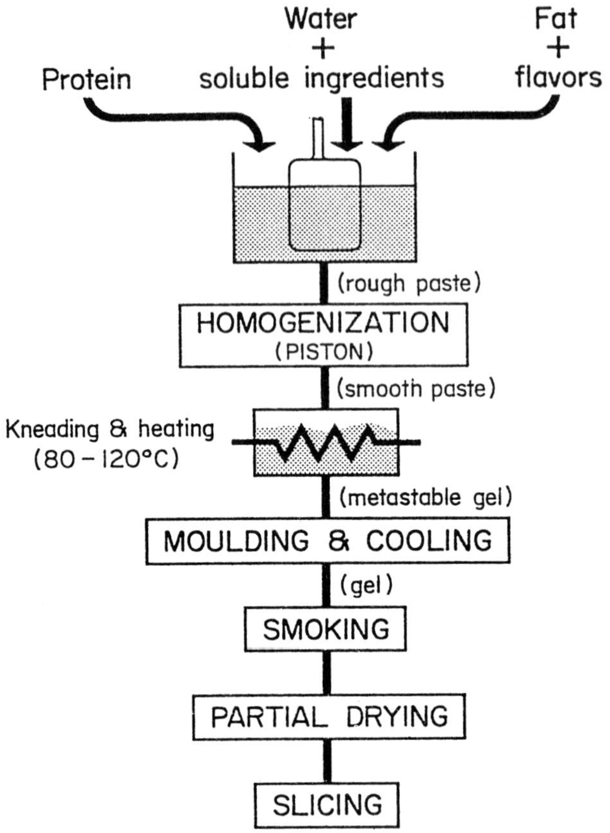

Figure 2. Gel-drying texturization.

molecules rather than the orientation of individual molecules: the texture will result from submacroscopic or macroscopic deformations of the hydrated protein material imposed by the process used. Textures thus obtained will always be stabilized sooner or later (i.e., during or after the texturization process) by a non-reversible coagulation or gelification of the protein constituents, as has been discussed formerly.

Cooking extrusion (2)

This process is based on the expansion of the protein material due to water vapour presure. The thermal stabilization of the structure takes place during and immediatley after the expansion phase. According to the characteristics of the extruder and its outlet head, the expansion process produces a lamellar structure including macroscopic fiber elongations.

Several other processes based on the same physical principle have been proposed; however, they do not always make use of an extruder; in some processes, dilation is even achieved by the expansion of a gas other than water vapour.

Table 2. Gel drying texturization (example of formulation)

		% solid matter
Promine D		60.6
Whey (dry)		10.1
Gelatine		3.03
NaCl		5.05
Coco fat		10.1
Glycerol		4.0
Sorbate		0.28
Fondagil	6	1.16
	10	1.16
	12	0.44
Colors	708	0.02
	752	0.02
	715	0.04
Caramel		0.4
$CaHPO_4$		0.8
Flavor		2.8
		100

The composition of the starting material will decisively influence the expansion phase and, consequently, the final texture of the product. Figure 1 shows the heat behaviour of a model protein mixture of soy/starch; the optimum corresponds to a ratio 8/20.

Gel texturization (3)

This process permits various technological applications from extrusion of gelled materials (strips, platelets, etc.) to production of stratified lumps by means of a contraction phenomenon achieved by partial drying or by osmotic processes. The protein materials used must be able to form a non-reversible (i.e. thermostable) gel. This property requires that the molecules of the protein material and of water be highly dispersed. Therefore, the preliminary hydratation of the material has a very important part in the process.

The gel texturization principle will be described by briefly discussing the process developed in our laboratory (Figure 2) [2]. The various components, including water, are thoroughly mixed and homogenized. Subsequently, the thin dough (Table 2) is subjected to a thermal treatment determining the formation of a 'progel' which has a definite shape after moulding and partial drying. The smoking and the drying process will lead to further internal tension, thus conferring on the product a particular texture resembling that of cured meat.

Tear texturization (4)

This technological approach can lead to various processes, a good example of its principle being the process proposed by the Southern Regional Research Center (USA) of New Orleans [1] for cotton protein: an aqueous suspension of cotton protein isolate is thoroughly stirred, its temperature being increased

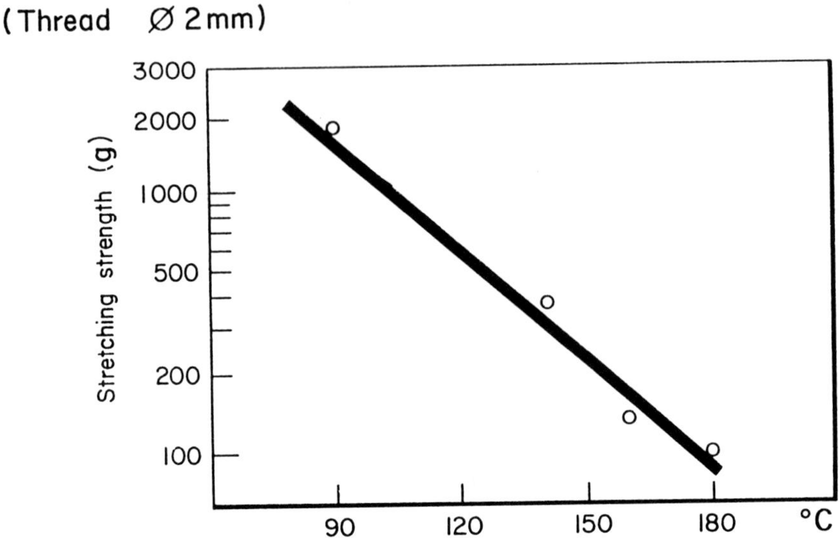

Figure 3. Extensometry of a protein isolate.

from 25°C to 90°C. The tearing effect occurring during coagulation produces a characteristic lamellar fiber-like structure. The difficulty with these kinds of processes has been in transposing the laboratory application to an industrial scale.

Melt spinning (6)

This texturization principle is based on the spinning of dry, or almost dry, proteinaceous material by thermal fusion (Figure 3) and has been studied in our laboratory. By blending the protein material with additives the melting temperature can be reduced, thus facilitating fiber production. This spinning process is simpler than wet spinning and is suitable for mixed materials such as protein concentrates. Being dry, the fibers obtained have a good shelf-life. They are hydrated before use.

Solvent texturization (7)

Several authors proposed to produce mixed polysaccharide/protein fibers without spinning. These processes include stirring a polar organic solvent (i.e. miscible with water), such as ethanol or acetone, with the aqueous solution of the chosen material.

The mechanism of fiber formation is very complex; it can be summed up as follows:

- Co-insolubilization of the two materials – protein and polysaccharide (e.g. of microbiological origin) – at least one of which has a linear structure (elongated chain) favorable to fiber formation.

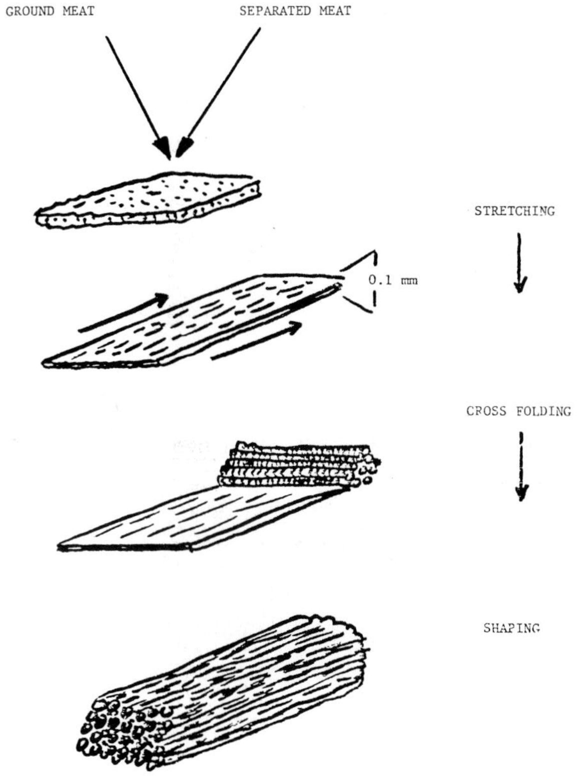

Figure 4. Stretch-cross-folding process (Battelle Patent).

– Furthermore, the formation of fibers depends on the viscosity of the aqueous solution that makes it possible, when mixed with the organic solvent, to produce fine streams of aqueous liquid which generate a fibrillar structure.

In our laboratory, we had the opportunity of testing such methods for texturizing protein/carbohydrate complexes which constitute the matter of unicellular organisms such as yeast. Contrary to the methods mentioned above, the polar solvent (and not water) was the carrier of the cellular materials.

Texturization by surface deformation (8)

Application of this process to material with a high protein content (e.g. protein concentrate) led to the industrial-scale development in our laboratory, of a technique called Stretch-Cross-Folding [3], the results of which are outstanding for certain formulations. The principle is based on the formation of thin layer ($<$ 0.1 mm) of hydrated and homogenized protein material. This highly stretched layer is released and cross-folded (Figures 4, 5, and 6).

Figure 5. Texture of a protein texturized by the stretch-cross-folding process (60 ×; scanning microscope).

An oriented structure, composed of a large number of small parallel fibers, is thus created and this constitutes the basis of the textured products. Stabilization of the structure is obtained by coagulation of the proteins by an appropriate thermal treatment. This process, for which a patent has been granted, is equally suitable for mixes of animal and vegetable proteins. The products thus obtained have various forms and sizes.

Texturization by freezing (9)

This procedure or approach has led to the development of several distinct processes. Only two of them will be mentioned.

The so-called process of *Freeze Texturization* [5] was developed by Battelle, the principle of which consists of imprinting a network of irregularly shaped, but parallel grooves on a half-frozen layer of ground meat or ground meat mixed with vegetable proteins. The final product is obtained by super imposing several of these striated layers and its appearance and texture are similar to those of an original meat product.

In such a process, the texture is due to the presence of longitudinal areas with a lower mechanical strength than the intermediate layers. Also, in this case, the quality of the texture depends on the coagulating and heat cohesion capabilities of the basic material.

Figure 6. Vegetable meat chunk made by the stretch-cross-folding process.

Several patents have been granted for another texturization principle by freezing: *cryoconcentration* of a hydrated paste of protein material. In this process, slow freezing just below 0°C leads to the formation of pure water crystal layers separated by unfrozen areas of concentrated proteins. Controlled refrigeration determines the formation of superimposed layers of protein material and ice; after heat coagulation of the proteins a textured lamellar-fibrillar product imitating the structure of meat is obtained.

'Biological' texturization

A texturization technique that differs fundamentally from the nine classes or types identified above is the production of meat substitutes on the basis of fibrillar mycelium of Fusarium graminearum [4]. This organism, when grown in a appropriate culture medium, produces a fibrous chitin-glucane matter embodying a protein material which is coagulated by heating. The application potential of this technique is promising as soon as some legal problems are resolved.

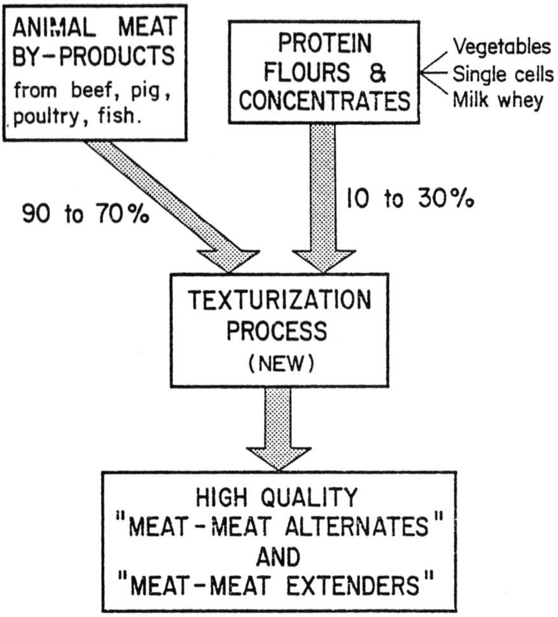

Figure 7. Upgrading of animal meat by-products by retexturization.

Future prospects

Several novel processes have been proposed to date, some of which are already being applied on an industrial scale despite their shortcomings. In our opinion, no 'ideal' process (i.e. which is economical, efficient and capable of supplying high quality texture) will be developed in the near future. In fact, success will be based essentially on a better control of the physico-chemical properties of protein materials suitable for texturization. Indeed, we need more cohesive and better orientable protein materials from the molecular as well as the macroscopic viewpoints. Furthermore, we have to study reticulation techniques (cross-linking) for proteins which meet nutritional requirements. Texturized food materials are, like plastics, complex polymers, the tridimensional conformation of which has to be organized and stabilized.

I would like to stress the priority of combining vegetable and animal proteins (Figure 7). Association of these two materials presents considerable advantages from the organolepitc, textural, nutritional, and economic standpoints. It is to be hoped that the vegetable proteins and the meat product industries will actively collaborate in order to carry out this project.

References

1. Cherry JP, Berardi LC (1981) Southern Regional Research Center (USDA), New Orleans, 182nd ACS National meeting, New York, August 1981. Division of Agricultural and Food Chemistry, paper 58

2. Giddey C, Rufer W. CH Patent No 542.593. Battelle, Geneva Research Centers
3. Giddey C, Rufer W. US Patent No 3,973.044. Battelle, Geneva Research Centers.
4. March RA (1981) RHM Research Ltd. The Lord Rank Research Centre, Lincoln Road, High Wycombe, Bucks, England. 182nd ACS National meeting, New York, August 1981. Division of Agricultural Food Chemistry paper no 54
5. Menzi R, Rufer W. CH Patent No 616.057. Battelle, Geneve Research Centers
6. Shimada, Matsushita J Agr Food Chem 28:413

Influence of processing on the nutritional value of proteins

P.A. FINOT

Nestle Products Technical Assistance Co. Ltd., Research Department, CH-1814 La Tour-de-Peilz, Switzerland

Key words: proteins, processing, nutritive value, chemical interactions, amino acids

Abstract. The nutritive quality of food proteins can be modified by the processes utilized for the preparation of foods. The effects of the physical and chemical treatments and the interactions between proteins and food components (proteins, sugars, oxygen, polyphenols, solvents, etc.) are discussed in the context of the utilization of new plant proteins for human consumption. Modifications of nitrogen digestibility, decreases in the amino acid availability, metabolic transit of the new compounds formed, and physiological effects are discussed.

Introduction

A certain number of processes are utilized for the preparation of traditional foods: cooking, baking, sterilization, concentration, drying, etc. Recently, new technologies have been developed to valorisate the vegetable proteins and to prepare new protein foods for humans and animals: extraction, isolation, detoxication, discolouration, texturization, etc. These food processes include:

- Physical treatments: milling, pH changes, heat-treatments, etc.;
- Chemical treatments: solvants, oxidizing agents, alkaline solutions, etc.;
- Biological treatments: enzymic hydrolysis, fermentation.

It is evident that chemicals intentionally added could react with proteins; actually, the food proteins can also react with the other food components through physical, chemical and enzymic interactions (Figure 1).

The different types of interactions of food proteins can be classified as follows: physical denaturation, modification in amino acid composition, destruction of the amino acids, protein–protein interactions, interactions between protein and oxidizing agents, protein–sugar interactions, protein–polyphenol interactions, protein–tannin interactions, protein–solvent interactions and others.

These physical and chemical modifications of the food proteins lead to changes in their nutritive quality (nitrogen digestibility and amino acid availability). The interaction products which are formed can induce in the

TREATMENTS AND INTERACTIONS	MODIFICATION IN FOOD PROTEIN
PHYSICAL TREATMENTS	
– fractionation (pH, leaching)	change in amino acid composition
– heat treatment	denaturation (enzymes, antinutritional factors) destruction of amino acids (desulfuration dehydration, deamidation) interaction with food components
CHEMICAL TREATMENTS	
– alkali (NaOH, NH_3)	amino acid destruction (cysteine, serine, arginine, racemization)
– oxidizing agents (H_2O_2, ClO Na)	oxidation of methionine, cyst(e)ine, tryptophan
– reducing agents (SO_2)	sulfocysteine
– solvents (chlorinated solvents)	reaction with cyst(e)ine
INTERACTIONS WITH FOOD COMPONENTS	
– proteins	isopeptides lysinoalanine–lanthionine
– oxidizing molecules (oxygen, hv, lipid oxidation, polyphenols, pigments)	oxidation of methionine, cyst(e)ine, tryptophan
– sugars	Maillard reaction
– polyphenols	reaction with lysine oxidation of methionine

Figure 1. Processing and modifications in food proteins.

living organisms specific physiological properties. The study of their absorption, excretion, transformation in the organism, retention, etc. is of great interest in understanding their nutritional properties and their physiological effects. This new approach, called 'metabolic transit', requires in most of the cases, the usage of isotopic techniques (Figure 2).

Protein denaturation

In their natural state, proteins have biological activities which are stable within narrow limits of pH and temperature: enzymatic activities (lipase, lipoxygenase, polyphenol oxidase, etc.) and antiphysiological properties (trypsin inhibitors, hemagglutinins).

The heat-treatments performed in the presence of water, which favour

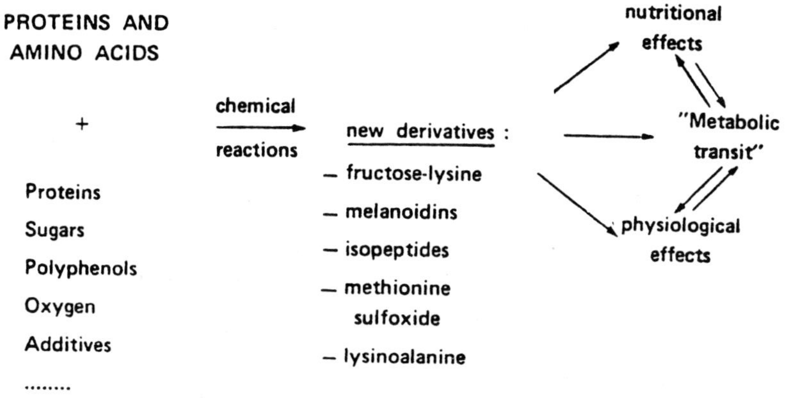

Figure 2. Importance of metabolic transit studies to explain the nutritional and physiological effects of the new molecules formed by interactions between proteins and other food components or additives.

protein denaturation, improve the digestibility of vegetable proteins. One can distinguish:

(a) a direct effect (or cooking effect) which destroys the cellular structure and denatures the proteins which are then more easily attacked by the proteolytic enzymes;

(b) an indirect effect due to the inactivation of the antiphysiological factors, particularly the trypsin inhibitors. These inhibitors need to be heat-treated in the presence of water to be inactivated. This is very well explained by Rackis [37] who found that the PER value of a soya meal increased gradually under atmospheric steaming in the same time as the trypsin inhibitor activity decreased. A long treatment (30 min) at 85–90 °C is more beneficial for inactivating the trypsin inhibitors and for increasing the PER of 'winged' beans than a short treatment (10 min at 120 °C) as shown in Table 1 [46]. The heat denaturation of the proteins also precludes subsequent problems of 'off flavours' and of the reduction of the nutritive quality of proteins, mediated by some enzymes (lipase, lipoxygenase, polyphenol oxidase) present in the plants.

Protein fractionation and changes in the amino acid composition

New technologies have been recently developed to fractionate plant proteins, in order to get fractions more or less rich in proteins (concentrates, isolates), to eliminate antinutritional factors, and to prepare protein fractions having specific functional properties.

The fractionations which are commonly utilized (leaching, solubility at different pH, precipitation at the isoelectric point) modify the composition

Table 1. Effect of two types of heat treatments on the nutritive quality of 'winged' bean [46]

'Winged' beans	P.E.R.[a]
Uncooked	< 0
10 min at 120 °C (autoclaved)	1.43
30 min at 85–90 °C	1.81
Casein (reference)	3.20

[a] Protein Efficiency Ratio.

Table 2. Essential amino acid content of soybean protein [48]

	Grams amino acid/16 grams nitrogen		
Amino acids	Defatted meal	Concentrate	Isolate
Lysine	6.9	6.6	5.7
Methionine	1.6	1.3	1.3
Cystine	1.6	1.6	1.0
Tryptophan	1.3	1.4	1.0

of the essential amino acids and particularly of methionine and cystine which are already limiting in the crude protein from plants.

According to Wolf [48] the content of some essential amino acids (lysine, methionine, cystine and tryptophan) decreases during the preparation of soya concentrates and isolates (Table 2). The more the protein is purified the lower the essential amino acids content. This is due to the loss of the more soluble protein fractions which are rich in methionine, cystine and lysine.

Another example is given by Hackler et al. [23] who found that the water soluble fraction of soya (soymilk) has a lower PER value (2.11) than the insoluble fraction (2.71) which is discarded as a by-product. The protein fractions obtained in a specific process should therefore be checked, especially if they are to be used as a quantitatively important ingredient of a protein food.

Destruction of amino acids

The chemical stability of the side chains of amino acids differs according to their structure. Heat treatments can induce the following modifications: deamidation of glutamine and asparagine, dehydration of N-terminal glutamine into pyrolidone-carboxylic acid [33] and transformation of tryptophan into mutagens during pyrolysis [40].

Alkaline treatments are responsible for the desulfuration of cyst(e)ine [44], deguanidination of arginine [50], dehydration of serine-(phosphate) [7, 19], and isomerization of all of the amino acids. The rate of isomerization is different from one amino acid to another; the most sensitive ones are

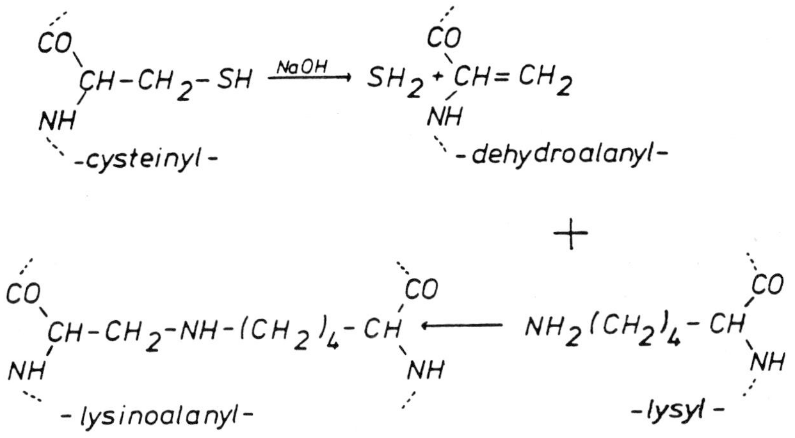

Figure 3. Formation of lysinoalanine.

aspartic acid, phenylalanine, and glutamic acid [32]. Some of these modifications are very important nutritionally, particularly the destruction of cyst(e)ine and the racemization of the amino acids.

Protein–protein interactions

Protein–protein interactions lead to the formation of several types of intra- or interchain cross-links: lysinolanine, lanthionine, ornithinoalanine and the isopeptides.

Lysinoalanine

Formation and processing. Lysinoalanine is formed by reaction of the epsilon-aminogroup of lysine with dehydroalanine (Figure 3); this last molecule results from the desulfuration of cyst(e)ine [7], the dehydration of serine and the dephosphorylation of serine-phosphate in casein [19], under severe alkaline or heat treatments [42]. The loss of cyst(e)ine begins at pH 10.0 and increases with time and temperature [12]; this loss is followed by the subsequent formation of lysinoalanine and other related substances, lanthionine [25], beta-aminoalanine [1] and ornithinoalanine [50].

Nutritional effects. The presence of lysinoalanine in a food protein is the consequence of the chemical reaction between two essential amino acids with the subsequent decrease in their biological availability. However, the loss of cyst(e)ine is much more important nutritionally than that of lysine since (a) lysinoalanine is only one of the products formed from the destruction of cyst(e)ine and (b) cyst(e)ine, as a sulfur containing amino acid, is already limiting in most plant proteins except cereals. Lysinoalanine forms cross-links which also contribute to the decrease in the nitrogen digestibility [11].

Metabolic transit. Lysinoalanine, when it is bound to protein, is partly absorbed by the gut and excreted in the urine [15]. Metabolic transit studies performed with radioactive lysinoalanine, labelled with ^{14}C in the lysine moiety, showed that (a) about 60–65% of the radioactivity excreted in the urine is present as lysinoalanine, (b) the rest of the urinary radioactivity is due to catabolites, and (c) the kidneys are able to split lysinoalanine and generate lysine which is incorporated into the proteins of the tubular cells [15, 16, 43].

Physiological effects. Lysinoalanine was found to induce renal defects (called nephrocytomegaly) in rats fed for several weeks a diet containing free or protein-bound lysinoalanine [22, 49]. This nephrocytomegaly corresponds to an hypertrophy of the nuclei and the cells of the renal tubules (pars recta); it has only been observed in rats and never in other species (mice, hamsters, rabbits, dogs and monkeys) [11], it was also found to be reversible [43]. According to Feron et al. [13], the LD isomer would be more active than the LL isomer. Theoretically, the two isomers are present in equal quantity in alkali-treated proteins. The mechanism of induction of the nephrocytomegaly in rats is still unknown and the extrapolation to what happens in humans is very difficult. In any case, the formation of lysinoalanine in food protein has to be reduced as much as possible.

Lanthionine – ornithinoalanine

Lanthionine is formed by a reaction between dehydroalanine and cysteine. Ornithinoalanine is formed by reaction between dehydroalanine and ornithine. Their formation contributes to a decrease in cyst(e)ine availability or in nitrogen digestibility. No harmful effect has been attributed so far to these two new amino acids. Their metabolic transit is not yet known.

Isopeptides

Formation and nutritional effects. During severe heat treatments, lysine can react with glutaminyl and asparaginyl residues to give isopeptide bounds, respectively ϵ-(γ-glutamyl)-lysine and ϵ-(β-aspartyl)-lysine [6] (Figure 4). The first one is biologically available while the second one is not, when given as free synthetic products [17]. It seems that when it is bound to proteins, the glutamyl isopeptide would be absorbed and therefore biologically available [26].

Metabolic transit. It is well established that the glutamyl isopeptide is absorbed by the gut and passes into the blood without hydrolysis [38, 47]. The kidneys are able to liberate lysine from isopeptides, like from lysinolanine, but

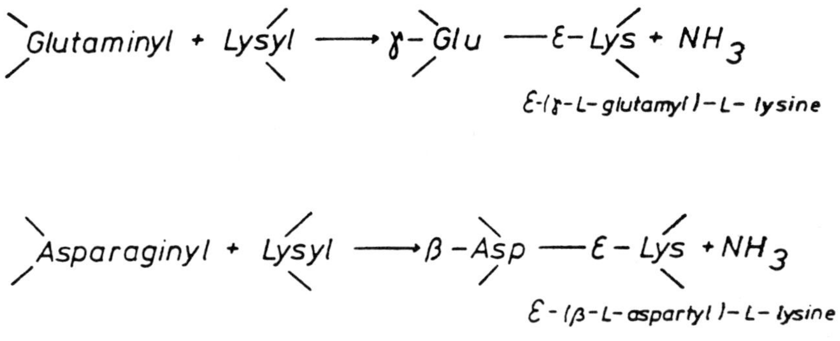

Figure 4. Formation of isopeptide cross-links.

in this case, this liberation is quantitative. The physiological effects of the isopeptides are not known.

Interactions between protein and oxidizing agents

Food proteins are sometimes in contact with oxidizing agents voluntarily utilized or involuntarily formed. Hydrogen peroxide is utilized for its bactericide properties, for the discolouration of food extract [41] and to detoxify aflatoxin [39]. Sodium hypochlorite has been proposed to detoxify aflatoxin in peanuts [36]. During the lipid autoxidation, free radicals and hydroperoxides are formed; photooxidation reactions catalysed by natural pigments (chlorophylle, riboflavin) [29] and acetone [20] have been described; polyphenols chemically or enzymatically oxidized into quinones also have oxidizing properties [45]. Three amino acids can be oxidized on their side chain: methionine, cyst(e)ine and tryptophan.

Methionine

Methionine can be oxidized into methionine sulfoxide and methionine sulfone (Figure 5). Methionine sulfone, which is rarely present in food proteins, was not found to be utilized as a source of methionine by the rat [5]. Methionine sulfoxide, which is easily formed in food proteins, was found to be fully utilized for growth and protein synthesis when given as a pure synthetic product [4, 16, 21, 34] and nearly completely utilized when given protein-bound [8] (Table 3). Metabolic studies showed that methionine sulfoxide is rapidly reduced into methionine [16] in the liver and kidneys by enzymes whose activity is enhanced by addition of NADH [2]. Although the presence of methionine sulfoxide in a food protein does not cause important nutritional modifications, it induces some physiological changes. In animals fed several weeks on a diet rich in methionine sulfoxide, the amino acid accumulates in the muscles at the same time as methionine

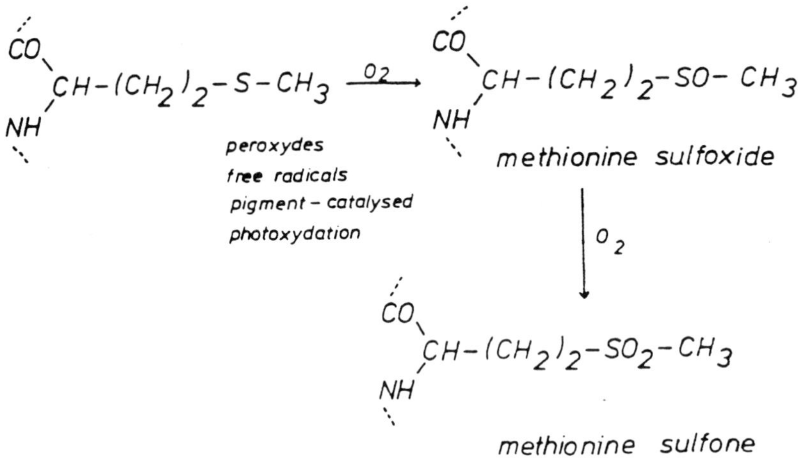

Figure 5. Oxidation of methionine.

Table 3. Nutritional effects of protein-bound methionine sulfoxide [8] [a]

	Casein	Oxidized casein	Statistical significance
PER	3.2	2.8	< 0.01
NPU	0.59	0.50	< 0.001

[a]PER = Protein Efficiency Ratio; NPU = Net Protein Utilization Value.

Table 4. Effect of a diet rich in methionine sulfoxide on the level of free methionine in muscle [8] [a]

Aminoacids	Micromoles/100 g of muscle		
	Casein	Oxidized casein	Oxidized casein + methionine
Methionine	2	14	17
Methionine–sulfoxide	12	245	245

[a]Diet: 13% protein for 25 days.

which is found at a level 15–20 times higher than the normal value [8] (Table 4).

Cyst(e)ine

According to Bennett [3], six different oxidation products of cystine and cysteine exist. Like for methionine, the less oxidized products would be utilized while the more oxidized ones would not be utilized. Their presence in food proteins is difficult to detect by conventional analytical methods.

Cystine–monosulfoxide	R – SO – S – R
Cystine–monosulfone	R – S – SO_2 – R
Cysteine–sulfinic acid	R – SO_2H
Cysteine–sulfonic acid	R – SO_3H

Figure 6. Oxidation of cyst(e)ine as detected by NMR spectroscopy [14].

Recently, Finley et al [14], using the NMR spectroscopy method, detected in model systems only four oxidation products: cystine monosulfoxide, cystine disulfone, cysteine-sulfinic acid and cysteic acid (Figure 6).

Tryptophan

Tryptophan can also be oxidized when treated with oxidizing agents; several oxidation products have been described: N-formyl-kynurenine, oxindol-alanine, β-carboline, etc. [24]. When treated with hydrogen peroxide, tryptophan is oxidized much slower than methionine [9].

Protein–sugar interactions

The interactions between protein and sugars, also called Maillard reactions, or non-enzymic browning reactions, are the most frequent in foods. These interactions occur between free amino acids or the epsilon-amino group of lysine in proteins and sugars. In the context of the plant proteins which do not contain large quantities of reducing sugars, this type of interaction seems not to be important; however, the plant proteins are present in seeds which also contain sucrose. Under certain conditions of treatment, sucrose is hydrolyzed into glucose and fructose which can therefore react in Maillard type reactions. Plant protein isolates or concentrates can also be used as ingredients for foods in which Maillard type reactions normally occur. The Maillard reaction produces a stable derivative between the free amino group of lysine and the reducing sugar involved; this product called Amadori compound (or fructose-lysine) is rather stable and its formation results in a loss of the biological availability of lysine. This step does not develop any browning.

The 'advanced' Maillard reaction proceeds from the decomposition products of the Amadori compound of lysine and produces reactive derivatives of the sugar moiety (secondary products also called premelanoidins) which

"Early" Maillard reaction | "Advanced" Maillard reaction

CHEMISTRY

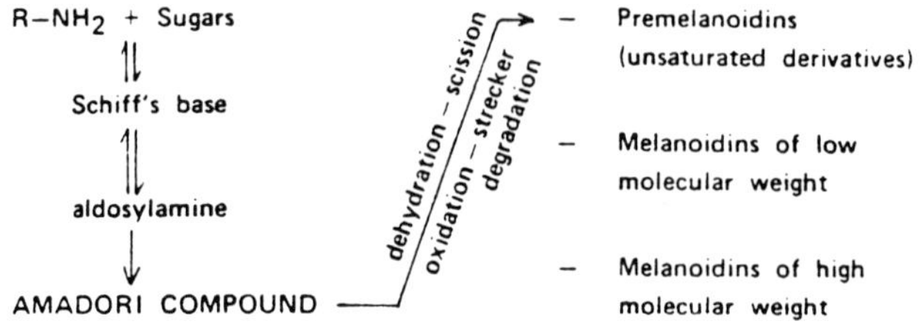

NUTRITIONAL AND PHYSIOLOGICAL EFFECTS

"Early" Maillard reaction:

- Loss of available lysine
- Inhibition of intestinal amino acid transportation

"Advanced" Maillard reaction:

- Loss of available lysine, methionine, tryptophan
- Decrease in nitrogen digestibility
- Inhibition of intestinal enzymes
- Change in plasma transaminases
- Hypertrophy of liver, kidneys and caecum

QUANTITY ABSORBED AND EXCRETED IN THE URINES

- protein bound fructosyl-lysine: 9 – 13 %

- premelanoidins: 26.9 ± 4.3 %
- melanoidins: 1.0 ± 0.1 %

Figure 7. The Maillard reaction: chemistry, nutritional and physiological effects, and metabolic transit.

polymerize into known pigments. These reactions occur during the sterilization process, the drying at high temperature and the storage under bad conditions (water activities between 0.4 and 0.6, high temperature).

The nutritional and physiological effects of the Maillard products can be summarized as follows:

- Amadori compounds: the Amadori compound of lysine is not biologically available; it is only partially absorbed by the gut (10–15%) and that absorbed is excreted in the urine without biochemical transformation.
- Premelanoidins: they are partially absorbed by the gut and the part which is absorbed is metabolized by the tissues; they have antinutritional properties, as described in Figure 7.
- Melanoidins: the melanoidins of high molecular weight are not absorbed

Polyphenols (caffeic acid, chlorogenic acid) + O_2 —— chemical or enzymatic oxidation ——→ Quinones

Quinones alone ——→ polymerization

Quinones + lysine ——→ addition compound + polymerized quinones

+ tryptophan ——→ oxidation or addition compound

+ methionine cyst(e)ine ——→ oxidation

Figure 8. Oxidation of polyphenols and reaction with the side chain of some amino acids.

by the gut. Their physiological effects are weaker than those of premelanoidins.

Protein–polyphenol interactions

Polyphenols (such as caffeic acid or chlorogenic acid) are present in various amounts in plants. In the presence of oxygen, they easily oxidize into quinones, either chemically (at alkaline pH) or enzymatically at pH 7.0 in the presence of polyphenol oxidase. Quinones easily polymerize into brown pigments and can react with or catalize the oxidation of the side chains of some amino acids (Figure 8). These reactions can occur during the extraction or the preparation of protein fractions from materials rich in polyphenols such as leaf proteins or sunflower seeds which contain 1.5 to 2.0% chlorogenic acid.

The reaction between the lysine residues and chlorogenic acid, as well as the nutritional effects of this reaction have been recently studied [28] on model systems using casein. In the absence of polyphenol oxidase, a drastic decrease in lysine availability occurs at pH 10.0 and above. In the presence of polyphenoxidase, the maximum decrease in lysine availability occurs at pH 7.0, i.e. at the pH of the maximum enzymic activity. These conditions are often observed in the technology used for the preparation of protein fractions (grinding, alkaline pH ...). However, under the conditions when the enzyme is active, the reaction is less rapid than under the alkaline conditions [28].

Animal assays showed a significant decrease of available lysine and a small decrease of available tryptophan and methionine (Figure 9) [27]. A

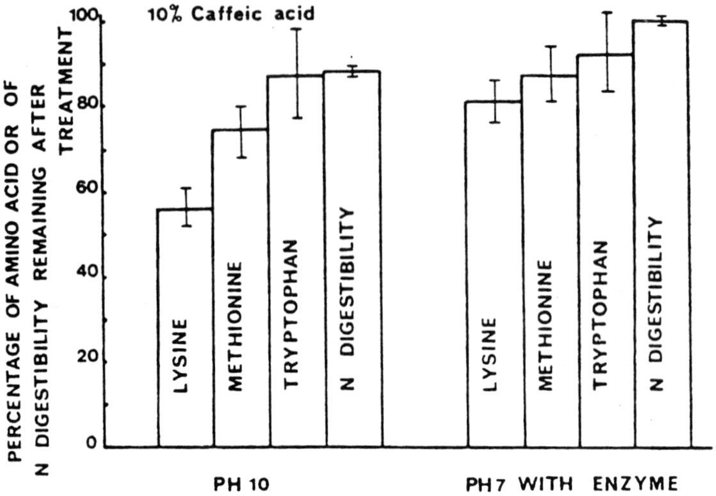

Figure 9. Available lysine, tryptophan and methionine in casein treated with cafeic acid at pH 7.0 in the presence of polyphenol oxidase and at pH 10.0 [28].

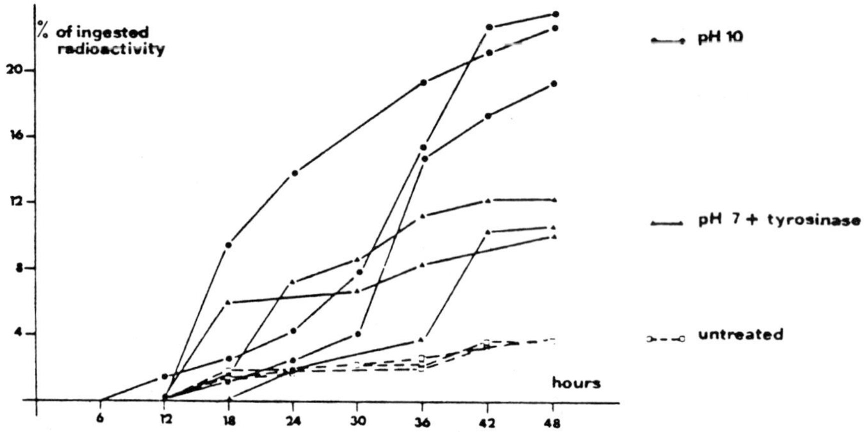

Figure 10. Fecal excretion of casein-bound tritiated lysine after treatment of casein biologically labelled with tritiated lysine, with caffeic acid at pH 10.0 and at pH 7.0 in the presence of polyphenol oxidase [28].

metabolic transit study performed with goat's casein biologically labelled with tritiated lysine showed that the lysine residues bound to caffeic acid were not absorbed by the gut (Figure 10) [27].

Protein–solvent interactions

Chlorinated solvents have been utilized to extract the oil from oil seeds. Trichloroethylene can react with the sulfhydryl group of cysteine to form

S-(dichlorovinyl)-cysteine, a toxic amino acid which produces an aplastic anemia in calves [30, 31]. 1,2-Dichloroethane is also able to react with cysteinyl residues to form 5,5′-ethylenebiscysteine [35].

Conclusions

During the extraction of plant proteins, as well as during the preparation and the storage of foods, some physical and chemical phenomena can occur which can modify the nutritive quality of proteins or induce certain physiological effects. These interactions have to be controlled by controlling the parameters involved in these modifications, development of analytical methods to quantify the modifications, evaluations of the nutritional consequences, and continued study of the physiological effects.

Some processes must be abandoned (chlorinated solvents, alkaline treatments); oxidizing agents have to be carefully manipulated; the conditions developing the Maillard reaction have to be controlled. The chemical and the biological methods exist to assess these modifications. Control of the raw materials, of the processes used and of the final products would normally be done in the developing centers and those industries which want to produce foods containing new protein sources.

References

1. Asquith RS, Booth AK, Skinner JD (1969) The formation of basic amino acids on treatments of proteins with alkali. Biochim Biophys Acta 181:164–170
2. Aymard C, Seyer L, Cheftel JC (1979) Enzymatic reduction of methionine sulfoxide. In vitro experiments with rat liver and kidney. Agric Biol Chem 43:1869–1872
3. Bennett MA (1937) Metabolism of sulfur. V: The replaceability of L-cystine in the diet of rats with some partially oxidized derivatives, Biochem J 31:962–965
4. Bennett MA (1939) The metabolism of sulfur. X: the replaceability of DL-methionine in the diet of albino rats with its partially oxidized derivative DL-methionine sulfoxide. Biochem J 33:1794–1797
5. Bennett MA (1941) The replaceability of dl-methionine in the diet of the albino rat with dl-methionine sulfone and dl-methionine methylsulfonium chloride. J Biol Chem 141:573–578
6. Bjarnason J, Carpenter KJ (1970) Metabolism of heat damage in proteins. Part 2: chemical changes in pure proteins. Brit J Nutr 24:313–329
7. Bohak Z (1964) Nϵ-(DL-2-amino-2-carboxyethyl)-L-lysine: a new amino acid formed on alkaline treatment of protein. J Biol Chem 239:2878–2887
8. Cuq JL, Besancon P, Chartier L, Cheftel JC (1978) Oxidation of methionine residues of food proteins and nutritional availability of protein bound methionine sulfoxide. Fd Chem 3:85–102
9. Cuq JL (1981) Modifications chimiques d'acides aminés indispensables sous l'influence traitements technologiques alimentaires. Incidences nutritionnelles et métaboliques. Ph D thesis. Université des Sciences et techniques du Languedoc, Montpellier, France
10. De Groot AP, Slump P (1969) Effects of severe alkali treatments of proteins on amino acid composition and nutritive value. J. Nutr 98:45–56
11. De Groot AP, Slump P, Feron VJ, Van Beek L (1976) Effects of alkali treated

proteins: feeding studies with free protein-bound lysinoalanine in rats and other animals. J Nutr 106:1527–1538
12. De Rham O, Van de Rovaart P, Bujard E, Mottu F, Hidalgo J (1977) Fortification of soy protein with cheese whey protein and the effect of alkaline pH. Cereal Chem 54:238–245
13. Feron VJ, Van Beek L, Slump P, Beems RB (1977) Toxicological aspects of alkali treatments of food proteins. In: Adler Nissen J (ed), Biochemical Aspects of New Protein Foods. Oxford: Pergamon Press, pp 139–147
14. Finley JW, Wheeler EL, Witt SC (1981) Oxidation of glutathione by hydrogen peroxide and other oxidizing agents. J Agric Food Chem 29:404–407
15. Finot PA, Bujard E, Arnaud M (1977) Metabolic transit of lysinoalanine bound to proteins and of free radioactive ^{14}C-lysinoalanine. In: Protein Crosslinking. Nutritional and Medical Consequences. Vol 86 B, Friedman M (ed), New York: Plenum Publishing Corporation, pp 51–71
16. Finot PA, Magnenat E, Mottu F, Bujard E (1978) Disponibilité biologique et transit métabolique des acides aminés modifiés par les traitements technologiques. Ann Nutr Alim 32:325–338
17. Finot PA, Mottu F, Bujard E, Mauron J (1978) N-Substituted lysines as source of lysine in nutrition. In: Friedman M (ed), Nutritional Improvement of Food and Feed Proteins. New York: Plenum Press, pp 549–570
18. Finot PA, Magnenat E (1981) Metabolic transit of early and advanced Maillard products. Prog Fd Nutr Sci Vol 5, pp 193–207. Pergamon Press Ltd
19. Freimuth U, Notzold W, Krause W (1980) On the alkaline treatment of proteins. Part V. The formation of lysinoalanine in β-casein and β-lactoglobulin. Die Nahung 24:351–357
20. Gennari G, Jori G (1970) Acetone sensitized anaerobic photooxidation of methionine. FEBS Lett. 10:129–131
21. Gjoen AU, Njaa LR (1977) Methionine sulfoxide as a source of sulfur containing amino acid for the young rat. Brit J Nutr 37:93–105
22. Gould DH, McGregor JT (1977) Biological effects of alkali-treated protein and lysinoalanine: an overview. In: Friedman M (ed), Protein Crosslinking. Nutritional and Medical consequences. Part B. New York: Plenum Press, pp 29–48
23. Hackler LR, Hand DR, Steinkrauss KA, Van Buren JP (1963) A comparison of the nutritional value of proteins from several soybean fractions. J Nutr, 80:205–210
24. Holt LA, Milligan B, Rivett DE, Stewart FHC (1977) The photodecomposition of tryptophan peptides. Biochem Biophys Acta 499:131–138
25. Horn MJ, Jones DB, Ringel SJ (1941) Isolation of a new sulfur containing amino acid (lanthionine) from sodium carbonate treated wool. J Biol Chem 138:141–149
26. Hurrell RF, Carpenter KJ, Sinclair WJ, Otterburn MS, Asquith RS (1976) Mechanism of heat damage in proteins. Part 7: the significance of lysine-containing isopeptides and of lanthionine in heated proteins. Brit J Nutr 35:383–395
27. Hurrell RF, Finot PA, Jaussan V, Cuq JL (1981) Nutritional consequences of protein polyphenol browning reactions. XIIth International Congress of Nutrition. San Diego, Abstract No. 260
28. Hurrell RF, Finot PA, Cuq JL (1982) Protein polyphenol reactions. Nutritional and metabolic consequences of the reaction between oxidised caffeic acid and the lysine residues of casein. Brit J Nutr 47:191–211
29. Jori G, Galiazo G, Tamburro AM, Scoffone E (1970) Dye-sensitized photooxidation as a tool for determining the degree of exposure of amino acid residues in proteins. The methionyl residues in ribonuclease A.J Biol Chem 245:3375–3383
30. McKinney LL, Weakley FB, Eldridge AC, Campbell RE, Cowan JC, Picken JC, Biester ME (1957) S-(dichlorovinyl)-L-cysteine: an agent causing fatal aplastic anemia in calves. J Am Chem Soc 79:3932–3934
31. McKinney LL, Picken JC, Weakley FB, Eldridge AC, Campbell RE, Cowan JC, Biester HE (1959) Possible toxic factor of trichloroethylene extracted soybean oil meal. J Am Chem Soc 81:909–915
32. Masters PM, Friedman M (1979) Racemization of amino acids in alkali-treated food proteins. Agr Food Chem 27:507–511

33. Mauron J (1972) Influence of industrial and household handling on food protein quality. In: Bigwood EJ (ed), Protein and Amino Functions. Vol 11. Oxford: Pergamon Press, pp 417–473
34. Miller SA, Samuel P (1968) Methionine sparing compounds. Proc Nutr Soc 27: 21A
35. Morrison AB, Munro IC (1965) Factors influencing the nutritional value of fish flour. IV Reaction between 1,2-dichloroethane and protein. Can J Biochem 43: 33–40
36. Natarajan UR, Rhee KC, Cater CM, Mattil KF (1975) Destruction of aflatoxins in peanut isolates by sodium hypochlorite. J Am Oil Chem Soc 52:160–163
37. Rackis JJ (1974) Biological and physiological factors in soybeans. J Am Oil Chem Soc, 51:161A–174A
38. Raczynsky G, Snochowski M, Buraczewki S (1975) Metabolism of ϵ-(γ-L-glutamyl)-L-lysine in the rat. Brit J Nutr 34:291–196
39. Rhee KC, Natarajan KR, Cater CM, Mattil KF (1977) Processing edible peanut protein concentrates and isolates to inactivate aflatoxins. J Am Oil Chem Soc 54:245A–249A
40. Sugimura T, Kawachi T, Nagao M, Yahagi T, Seino Y, Okamoto T, Shudo K, Kosuge T, Tsuji K, Wakabayashi K, Ittaka Y, Atai A (1977) Mutagenic principle(s) in tryptophan and phenylalanine pyrolysis products. Proc Japan Acad 53:58–61
41. Staron T (1977) Une méthode d'obtention des proteines végétales pour l'alimentation humaine et animale. Indust Alim Agric 11, 1197–1217
42. Sternberg M, Kim CY, Schwende FJ (1975) Lysinoalanine: presence of foods and food ingredients. Science 190:992–994
43. Struthers BJ, Dahlgren RR, Hopkins DT, Raymond ML (1979) Lysinoalanine: Biological effects and significance. In: Wilcke HL, Hopkins DT, Waggler DH (eds), Soy Protein and Human Nutrition, London: Academic Press, pp 235–260
44. Swan JM (1957) Mechanism of alkaline degradation of cystine residues in proteins. Nature 179:965
45. Synge RLM (1975) Interactions of polyphenols with proteins in plants and plant products. Qual Plant 24:337–350
46. Temler R, De Rham O (1979) Unpublished results
47. Waibel PE, Carpenter KJ (1972) Mechanism of heat damage in proteins. Part 3: studies with ϵ-(γ-L-glutamyl)-L-lysine. Brit J Nutr 27:509–515
48. Wolf WJ (1970) Soybean protein: their functional, chemical and physical properties. J Agr Food Chem 18:969–976
49. Woodard JC, Short DD (1973) Toxicity of alkali-treated soy-protein in rats. J Nutr 103:569–574
50. Ziegler K, Melchert I, Luerken C (1967) N^{ϵ}-(2-amino-2-carboxyethyl)-ornithine, a new amino acid from alkali treated proteins. Nature 214:404–405

Natural antinutritional factors present in European plant proteins

R. FERRANDO

107 rue de Reuilly Bat. 1, 75012 Paris, France

Key words: antinutritional factors, anti-enzymes, anti-hormones, hemagglutinins heterosides, toxic amino acids

Abstract. The great variety of agricultural productions is only one form of the various manifestations of nature. It is a current error today to believe that all that is natural is perfect, and, from a nutritional point of view, devoid of danger. However, numerous examples such as toxic mushrooms and fishes prove the contrary. Protein plants are not an exception to the rule. Although the side effects of some of their componants are not always as severe as those of mycotoxines or ichyotoxines, they nevertheless do exist. The list of noxious compounds present in legume seeds or other plant seeds is rather long. It ranges from cyanogenetic heterosides and toxic amino-acids, liable to lead to death, to the antienzymes, hemagglutinins, antivitamins and flatulence producing substances. Also, included are mineral chelating substances, especially phytic acid, oestrogenic hormones and anti-hormones. Some populations show a genetic sensitivity of food such as beans, leading to the terrible favism syndrome. The existence of pharmacogenetic problems in human nutrition has not been considered enough. In animal nutrition they are of less importance because of the possibilities offered by animal genetics.

Clinicians and nutritionists know the existence of these substances. Repeated experiments clearly indicate their more or less marked toxicity. Agricultural and technological means are used to surpass and/or reduce these toxical or pharmacodynamical actions. In many circumstances, re-adjustments of the dietary balances are very useful. Thus, there are antagonisms between biology and the economy of animal production, interrelated with human nutrition. The art of animal production is to reduce if not to suppress these antagonisms without taking into consideration the intelluctual deviations and aberrations of those who are forgetting more and more the imperatives of agriculture, whose adaptation is primordial to prevent the world from dying of starvation.

Introduction

Many people are inclined to believe that all which is natural is perfect. Nutritionists know that this is not true. This so-called ‘perfection’ has a lot of failures due to the presence of substances whose number, distribution and properties are becoming more and more well-known. Protein plants, belonging to various botanical families whose seeds are particularly protein rich, are not an exception to the rule. In the present paper we shall examine the following factors, actions or properties of protein plants cultivated in Europe:

- noxious heterosides,
- toxic amino acids,

Table 1. Main heterosides of protein-rich seeds or of their by-products

Source of seed containing the heteroside	name of the heteroside	Products from hydrolysis
Various cruciferae including rapeseed (*Brassica napus oleifera*)	Glucosinolate (progoitrine)	D. glucose + 5-vinyl-2-thioxazolidone (goitrine) + H_2SO_4
Vetches (*Vicia angustifolia et sativa*)	Vicianine	Vicianose + HCN + benzaldéhyde
Lima beans (*Phaseolus lunatus*)	Phaseolunatine	D. glucose + HCN + Acetone
Chick peas (*Cicer arietinum*) and broad beans (*Vicia faba*)	Phaseolunatine	HCN and various compounds, undetermined reaction
Trefoil (*Lotus arabicus*)	Lotusine	D. glucose + HCN + lotoflavine
Linseeds (*Linum usitatissimum*) and linseed cakes	Linamaroside	D. glucose + HCN + Acetone

– anti-enzymes, hemagglutinins,
– hormonal and antihormonal actions,
– mineral chelating properties,
– miscellaneous properties.

Contrary to the situation extant for the most part in the field of animal nutrition, we shall mention the existence of genetic sensitivities of certain human populations to some of these seeds. In fact, these pharmacogenetic problems may to a certain extent be compared with individual food allergies caused by the most unexpected types of food [10]. After having discussed those problems we shall conclude by discussing the possibilities of reducing, if not supressing completely, the consequences of the natural antinutritional factors in these European sources of plant protein (PAG, Protein Advisory Group of the United Nations System, 1973 [36]).

Noxious heterosides

These heterosides are not limited to seeds. Young sorghum shoots may, for instance, contain Dhurrine; the hydrolysis of which gives: D glucose + HCN + p-hydroxybenzaldehyde. Various authors have made an overall survey of that question [19, 22, 30, 41]. (The latter reports include bibliographical data concerning not only heterosides, but also all of the substances which will be treated in this paper.)

Table 1 shows the main heterosides and their hydrolysis products; their antinutritional actions are variable. Release of HCN may cause fatal injuries while goitrine may bring about changes in the thyroid gland. Later, when examining the hormonal and antihormonal effects, we shall discuss the consequences of these heterosides on the endocrine system.

Some exotic bean varieties may supply more than 300 mg HCN per 100 g. Prolonged cooking, intended to eliminate the toxicity by destroying enzymes which hydrolyze the heterosides, is not always efficient. Individuals who have consumed Lima beans treated in this way may still eliminate traces of HCN in their urines. It has been assumed that certain digestive enzymes or enzymes secreted by enterobacteria such as *Escherichia coli,* may also play a part in the hydrolysis of heterosides.

Toxic amino acids

Beside the amino acids which are essential to the maintenance and perpetuation of life and which in some instances, especially when supplied in excess, may be toxic, there are others which are toxic naturally. Among more well-known ones are those of legumes belonging to the genus *Lathyrus,* which may cause neurolathyrism which is characterized by neurological injuries. The following list was established by Somogyi [42] :

(1) L α, γ-diaminobutyric acid in *L. latifolius, sylvestris, aurantius* and 10 other *Lathyrus.*

(2) N γ, oxalyl α, γ diaminobutyric acid in *L. multiflora, undulatus* and 8 other *Lathyrus.*

(3) N β, oxalyl α, β diaminopropionic acid in 14 *Lathyrus,* especially in *L. sativus, clymenum, latifolius.*

(4) β-oxalylaminoalanine in *L. sativus.* This seed also contains α-amino-ε-guanido caproic acid which is also present in *L. cicera* and *L. clymenum. L. sativus* also contains another toxic compound: N-β-glucopyranosyl-N-α-L-arabinosyl-α, β-diaminopropionitrile.

(5) β-cyanoalanine and γ-glutamyl-β-cyanoalanine are found in *Vicia sativa* and in 15 other vetch varieties. These legumes, which now exist for the most part outside of Europe, could be grown in Europe.

Osteolathyrism, due to *L. odoratus* and *L. pusillus,* causes severe disorders in bone and connective tissue development. In the latter, lesions of the aorta may occur. The responsible compound is β-N-γ-L glutamyl amino propionitrile [16]. The active fraction is β-amino-propionitrile: $NH_2-CH_2-CH_2-CN$. The toxic substance seems to inhibit the formation of elastic fibres in the aorta (due to an increase in fibroblast proliferation) which leads to an irregular elaboration of collagen fibres. The action seems to depend on an inhibition of a copper containing enzyme, amino-oxidase.

Various broad-bean varieties contain proportions of vicine ranging from 4.90 to 8.25 mg/g. This glucoside might be responsible for specific side effects in persons exhibiting erythrocyte glucose-6-phosphate dehydrogenase (G 6 P D) deficiency [12].

Anti-enzymes and hemagglutinins

These compounds are found not only in plants used as a source of protein, but in other plants as well. The negative side effects of feeding raw soybeans to rats was first reported by Osborne and Mendel [35]. Heating prevents these effects. Using raw soybeans, Kunitz [29] isolated a protein which combined with trypsin to form an inactive complex. These properties were first widely studied in soybeans, but thereafter it was established that other legume seeds, such as *Pisum sativum,* vetches, etc..., exhibited an inhibitory effect on growth in chicken [26]. As a matter of fact, trypsin inhibitors exist in all legume seeds [31] and often coexist with anti-α-amylases [40]. However, the anti-enzyme action varies from one legume seed to another. That of *Vicia faba* is 5-fold lower than that of soybeans or beans.

There is no correlation between the protein content of broad beans and their content of protease inhibitors. The aqueous extracts of the seeds also inhibit not only trypsin but also α-amylase and liase. Broad bean varieties with white flowers do not seem to contain any anti-enzymes [23]. These differences according to varieties in one and the same legume have also been observed in *Vicia faba, Vignia sinensis* and *Pisum sativum.* The concentration of anti-trypsin activity is 2–3 times higher in winter peas than in spring peas [46]. According to Bingefors [8] the same differences are recorded in peas cultivated in Sweden. This author also noted that, like in broad beans, the level of trypsin and tannic acid inhibitors are lower in pea varieties with white flowers.

Rackis, in a recent review, gives an overall survey of soybean trypsin inhibitors [39]. Many authors have attempted to explain the mechanisms of this inhibition which lead to pancreas hypertrophy. In the rat, this reaction seems to be the first one of the organism. It results from a hypersecretion of pancreatic enzyme and consequently from an increased expenditure of proteins, especially sulphur amino acids which are abundant in trypsin. This might explain the further rise in the requirement for methionine, the limiting factor of soybean proteins and of other legume seeds [31]. The reactions seem to be varied, at least with soybean meal and in the rat [13]. These authors [13] observed that feeding raw soybeans reduced the synthesis of amylase-like iso-enzymes and increased that of two trypsinogens and one chymotrypsinogen. They also assumed that there is an adaptation of the pancreatic secretion in the presence of trypsin inhibitors [14]. More recent explanations of these actions have been given. On account of the low nitrogen digestibility of beans (74.2% to 54.5% according to the age of the experimental subject) as compared to casein nitrogen (97.2% to 92.2%), emphasis has been laid on the massive renewal of the gastrointestinal mucosa during legume seed consumption. This is shown by the DNA content of the extreta [4]. A cytoxic effect of legumes on the epithelial cell of the small intestine has even been mentioned [43]. Differences in activity seem to exist; *Phaseolus acutifolius* being considered as more active than *Cicer arietinum.* This cytotoxic action might be the cause of a lower intestinal absorption. In our opinion since this activity accelerates the renewal of intestinal mucosa, it may also contribute to an increase in the requirement for retinol.

The action of hemagglutinins is similar to that of anti-enzymes. These proteins called phytohemagglutinins were discovered in 1953 [25]. It is assumed that 25% of the inhibition of growth induced in the rat by feeding raw soybean is due to these compounds. This hemagglutinating activity also exists in various beans, lentil and wheat. The hemagglutinating and antitrypsin activities of various foodstuffs were studied by Liener in 1973 [31].

Table 2. Hemagglutinating and antitryptic activities[a] of extracts of raw legume seeds[b]

Nature of the legume seed	Hemagglutinating activity (HU/ml)	Antitrypsin activity (TIU/ml)
Phaseolus vulgaris		
Black variety	2450	2050
White variety	3650	1552
Cicer arietinum	0	220
Cajanus Cajan	0	418
Phaseolus aureus	0	260

[a] 10 parts of finely ground meal were suspended in 90 parts of 1% NaCl solution, and then clarified by centrifugation; the clarified supernatant was used to determine hemagglutinating units (HU) and trysin inhibitor units (TIU) per ml.
[b] From Leiner [31].

Table 2 shows that there are large variations in the levels of these antinutritional substances from one legume to another. The nature and effects of the hemagglutinins vary from one seed to another (e.g., see data in Table 3).

In soybeans, the influence of hemagglutinins seems to be less marked than that of anti-enzymes [45]. The effects on rat growth of a soybean extract devoid of its hemagglutinins, but whose antitrypsin activity is maintained, do not differ much from those of unheated soybean meal as measured by rat assay (Protein Efficiency Ratio $\bar{x} \pm$ S.D.):

	Protein Efficiency Ratio
Extract without hemagglutinin	1.13 ± 0.24
Extract with hemagglutinin	0.91 ± 0.39
Extract with heated hemagglutinin	2.25 ± 0.51
Unheated raw soybeans	1.01 ± 0.21
Heated raw soybeans	2.30 ± 0.35

However, these findings are in contradiction with those reported in 1973 by Liener [31] citing his own results reported in 1962 in co-operation with Honovar and Shih [24]. Once again, we observe large differences from one species to another and sometimes from one variety to another. However, this statement cannot be generalized.

Table 3. Effect of the purified hemagglutinin fraction of black or white beans on the growth of rats[a]

Hemagglutinin source	Purified hemagglutinin in the diet	Mean weight (g/day)	Mortality level (100% mortality observed after indicated number of days
Phaseolus vulgaris Black variety	0.0	+ 2.51	Null
	0.5	+ 1.04	Null
	0.5[b]	+ 2.37	Null
	0.75	+ 0.20	Null
	1.2	− 0.91	15–19
	2.3	− 1.61	12–17
	4.6	− 1.72	5–7
Phaseolus vulgaris White variety	0.0	+ 2.31	Null
	0.5	− 0.60	13–16
	0.5[b]	+ 2.29	Null
	1.0	− 0.87	11–13
	1.5	− 1.22	4–7

[a]From Liener [31].
[b]The hemagglutinin solution was boiled for 30 min. This treatment destroyed completely the hemagglutinating activity. The dried coagulum was added to the diet at the specified levels.

These antinutritional compounds have not only a protecting effect on the seed embryo [22], and subsequently on whole seeds, acting as natural pesticides, but also play a very important nutritional role [44]. According to the botanical origin of the seeds, these compounds may reduce the protein efficiency of a given food to various extents by rendering the theoretically adequate nitrogen supply insufficient. This may increase the nitrogen needs beyond the level required by the physiological status of an organism and the conditions of the environment to which it lives.

Estimations of the ability of the new types of feeds such as protein plants to contribute towards satisfying the protein requirements of the organisms depend not only on their amino acid content, but also on the repercussion of anti-enzymes and other compounds on the digestibility of these amino acids. A chemical analysis, even performed very thoroughly, is not sufficient. It must be accompanied by further tests, and in particular by the measurement of the Protein Efficiency Ratio or even of the Biological Value of the protein. Only such an approach as this leads to an acceptable biological evaluation. We shall underline the importance of this problem in our conclusions and draw attention to the necessity of evaluating the positive and negative consquences of technological treatments of feeds.

Hormonal and anti-hormonal actions

It has been known for a long time that various plants contain phenol compounds with oestrogenic activity [7]. In particular, forages contain these compounds, but genistein has also been isolated from soybeans [7]. More recently, Drane et al. [15] detected oestrogen activity in 16 soybean samples, as measured on mouse uterus. Also, an acetone-alcohol extract of the meal of these seeds showed the same amount of activity due to the presence of daidzeine as was due to genisteine. Daidzeine, a phytohormone present in the form of 7-glucoside-daidzeine, is more potent than gesteine [15].

It would be interesting to examine systematically the oestrogen properties of all legume seeds. This does not seem to have been done except in the case of soybeans, chick peas, and beans [6]. We showed that the oestrogen activity of plants like wheat and carrots, expressed in DES equivalents, varies according to the type of fertilizer used [18]. This activity might also contribute to improved protein anabolism, as shown in lucerne and forages in general.

The anti-thyroidian actions of heterosides of cruciferae seeds was noted above. In an experiment using dairy cows, 25% rapeseed meal, with a high or low level of heterosides, were added to the concentrates. In both cases, a lowering of the milk iodine content and of the blood level of thyroxine was observed [37]. An effect on the thyroid of calves and children fed with milk from these cows was also observed. However, only the meal with the highest heteroside concentration reduced the blood level of thyroxine of the calves and modified the histological structure of their thyroid glands. Rats receiving the same milk showed the same phenomenon. Administration of an iodine supplement (61 g/day/rat) prevents hypertrophy of the thyroid gland.

As affected by rumen fluid, progoitrine leads to production of 5-vinyl-1, 3-thiozolidine-2-thione and, to a lesser extent, 5-vinyl-1,3-oxazolidine-2-thione as well as other components [38]. However, Japanese authors estimate that goitrine does not modify the nitrogen balance of adult rats [1].

Anti-thyroidian substances are present not only in cruciferae, but also in soybeans. In the review mentioned above, Liener [31] indicates various studies pointing out the goitrogen effects of unheated soybean in rats, chickens, and even children. This author emphasized, in agreement with other authors, that an addition of iodine counteracts the antithyroid effect of these seeds and of their by-products.

At present, rapeseed varieties with low glucosinolate concenttations are available which, when fed, lead to good growth performances [11]. Elimination of the heteroside also improves the feeding value of a rapeseed oil-meal [47]. It is also interesting to point out that a stabilization of the demographic development in some Tibetan populations has been observed for years. These populations consume large quantities of peas (*Pisum sativum*) containing m-xylohydroquinone which suggests that this protein plant might therefore have contraceptive properties. It would be interesting to continue those studies both from a fundamental and practical point of view.

Mineral chelating properties

Legume seeds or other plant seeds may contain organic compounds, such as phytic acid and oxalic acid, which may decrease the utilization of dietary minerals, macro- and trace elements. In this case, the values predicted by chemical analyses will be invalidated.

Phytic acid or inositolhexaphosphoric acid is an ester resulting from the combination of the six secondary functional groups of inositol (cyclohexanol) with six orthophosphoric acid molecules. Discovered in cereals at the end of the last century, this ester is the only natural compound containing one phosphorus atom per each carbon atom. It can form simple salts with a metal or complex salts with several metals, all within the same molecule. These phytates like phytic acid itself may be chemically hydrolysed (under heating) or enzymatically with phytase. The reaction leads to release of a vitamin (inositol) and six orthophosphoric acid molecules. Considering the same plant segment, the phytic acid content varies according to cultivation conditions and the botanical species studied. Relative to the total quantity of phosphorus, phytic acid phosphorus represents 40–99% of the total phosphorus in dry legume seeds and 25–60% of the total phosphorus in nuts.

The seed pericarp together with part of the adhering neighbour layer and the embryo contain 10–20 times more phytic acid than the endosperm. Table 4 shows the phytic acid phosphorus content of the main protein-rich seeds.

Some plants contain more or less large amounts of phytase. Indeed, the activity of this enzyme varies much from one species to another, but there is no correlation between this activity and the phytic acid content. The phytase activity of lentils is very high, but the proportion of phytic acid is low. The contrary is the case for soybean, i.e. low enzyme activity and a high phytic acid content.

Phytic phosphorus is used differently according to animal species. In cattle and sheep, the ruminal microorganisms hydrolyse almost all (or all) of the phytic acid, whereas in poultry, pigs and man the efficiency of utilization of phytic acid phosphorus is only 2–10% of that of mineral phosphorus. The presence of phytic acid phosphorus also seems to increase slightly the requirement for vitamin D_3 in chickens and perhaps in other species. It seems that this vitamin slightly improves the digestibility of phytic acid phosphorus.

The level of oxalic acid is not very high in legume seeds. Only *Dolichos biflorus* contains a rather large amount, i.e. 417 mg/100 g, and *Lathyrus sativus* contains the second highest level with 144 mg/100 g.

Like in the protein-rich plants, the percentage of phytic acid phosphorus is also high in cereals, ranging from 32–97% relative to total phosphorus.

Table 4. Phytic acid phosphorous content of protein-rich seeds

Scientific name	Current name	Total phosphorus (mg/100 gm)	Phytic acid phosphorus (mg/100 gm)	Phytic acid phosphorus (mg/100 mg total phosphorus)
Cicer arietinum	Chick peas	276–372	140–354	49–95
Phaseolus of various varieties	Broad beans	196–405	152–250	52–77
Faba vulgaris	Broad beans	527	500	95
Vigna catjang	Dolique beans	462 (447–478)	213 (167–258)	46 (37–54)
Dolichos lablab	Egyptian bean	417	248	59
Pisum sativum	Peas	319 (286–354)	117 (105–130)	37
Cajanus indicus	Pigeon pea	263	200	76
Lens esculenta and *culinaris*	Lentils	326–339	89–295	26–90
Soja hispida	Soyabean	614 (442–838)	402 (231–575)	65 (52–68)
Lathyrus sativa	Jarosse vetch	262	82	31
Nux	Nuts	500	120	24
Linum usitatissimum	Linseed	880 (860–900)	741 (606–785)	84

From Gontzea et al [22]

Miscellaneous properties

Referring to Bleumink [10], we mentioned the allergic effects of various plants in the first part of this paper. Pre-ruminant calves fed with heated soybean possess specific soybean glycinine and β-glycinine antibodies. These allergens seem to play a part in the development of various digestive disorders and in the appearance of serum antibodies in these calves [27]. In our opinion, no other studies are available on protein-rich seeds.

The anti-vitamins present in these seeds are the following:

(1) *Phaseolus vulgaris* – anti-vitamin E of unknown nature which, when consumed by sheep, leads to injuries (which are also observed in rats and chicken) [3].

(2) Various peas – anti-pantothenic acid or pisamine. Injected in rats, it reduced the level of pantothenic acid in the liver, kidney and brain; there seems also to be an adverse effect on vitamin B_{12} [34].

(3) Linseed meal, *Linum usitatissimum,* anti-vitamin B_6 or linatine producing l-amino-D-proline and glutamic acid. Linatine is 1 (N-γ-L-glutamyl-amino)-D-proline. It was isolated by Kratzner and Williams in 1948 (cited by Klosterman [28]. A supplement of pyridoxine should be added to the rations of pigs fed 300 g lineseed meal per kg diet [9].

(4) According to various authors, soybeans also contain antivitamin D_3,

Table 5. Effect of various supplements added to Boyd medium on gas production by Clostridium[a]

Supplement (mg per tube)	ml gas produced after incubation for	
	2 hours	3 hours
None	0	3.0
50 mg beans	+ +	6.7
50 mg bean ashes	+ +	6.7
50 mg soyabean ashes	2.0	14.0
15 mg yeast extract	4.5	48.1
Yeast extract + beans	18.0	75.0
Yeast extract + beans ashes	25.0	75.0
Yeast extract + soyabean meal ashes	25.3	75.0

[a]From McGinnis [33]

[31]. This perhaps only involves the action of phytic acid on the aforementioned requirement for cholecalcipherol.

Legume seeds contain flatulence-producing compounds. This has been empirically established a long time ago. There are individual variations in the production of intestinal gases, as shown in several studies [20]. The main components of these gases are CO_2, H, and sometimes CH_4. Oligosaccharides like raffinose and especially stachyose and verbascose seem to be responsible for this gas production in the case of *Phaseolus aureus, Phaseolus mungo* and *Lens culinaris* [21]. These studies also showed that addition of yeast extract to the diet stimulates the gas formation (Table 5). Bean and soybean ashes (Table 5) have the same effect [33].

Through selection it is, however, possible to obtain legume varieties which cause less gas production. It has also been established that some kinds of food or beverages such as coffee include anti-flatulating compounds such as caffeic, ferulic and chlorogenic acids.

General discussion and conclusions

All foodstuffs have not only positive effects, but also negative ones which are more or less bound to the imbalance of their nutrients or to the presence of anti-nutritional substances. As shown in this paper there are many types of those substances. Even the egg, which is a reference protein, contains anti-trypsin and avidine, an anti-vitamin (biotin chelating factor). In addition, some people are allergic to eggs.

It is absolutely necessary to be well acquainted with these various actions and, as mentioned above, to go beyond simple chemical analysis. The latter may inform us about a certain number of points, but not about undetermined factors which still remain partly obscure. Only studies made with laboratory animals and thereafter farm animals (kept in a normal nutritional balance and thus including practical aspects of the problem) are adequate for determining the real feeding value of a new protein source. A constant observation of living beings during such experiments and control of all their physiological activities remain one of the bases of nutrition and the best guarantee for searching, obtaining and preserving the nutritional balance' [17]. Some rules will be suggested for bioproteins by the Scientific Committee for Animal Nutrition within the E.E.C. In our opinion, they should be applied to protein-rich plants as well. A technique based on addition of cooked or raw white beans, *Phaseolus vulgaris,* to a standard diet allows for measuring, within 10 days, a reduction in net protein utilisation values. In that experiment, the author confirmed the presence of anti-nutritional factors in the seeds studied by Bender [5]. This technique seems to be valuable, but there is always need for other investigations or approaches as well.

When the negative effects of foodstuffs have been detected and quantitated more accurately, they may be reduced or even eliminated in several ways. First of all, we can live with our toxic substances. It is once again a question of the establishment of a new nutritional balance. This is the case when minerals are added to a basal diet containing too much phytic acid phosphorus. Likewise, it is possible to counteract the effect of anti-vitamins by providing an additional supply of the factors (vitamins) neutralized by these compounds.

However, other methods are available. Technology has provided us with many means. The action of heating on the destruction of anti-enzymes and hemagglutinins has been well established. But these treatments may destroy some amino acids and vitamins. For maintaining the nutritional value of food subjected to such treatments, it is necessary to check that heating temperature and length during industrial and culinary preparations reach and do not exceed the optimum required to eliminate the effects of inhibitors or phytohemagglutinins without altering the basic nutrients. Once again, the criterion of evaluation should be both biological and analytical.

Selection of varieties devoid of anti-nutritional substances [36] for those containing only small quantities might also be a solution. This method has been applied with success to rapeseed. However, in that case, too, the analytical results should be confirmed by large-scale biological evaluations with several species. Such studies may also sometimes contribute to the discovery of interesting pharmacological actions. The study of the anti-nutritional factors of protein-rich plants not only represents a merely technical problem, it also leads to considering the nutrition sciences and their application, i.e. feeding, from a more philosophical point of view. The notion of perfection always being imputed to all that is natural is very close to the myth of the 'good savage'. None of them survive an objective critical analysis. Even modifications intended to only improve the natural, may in turn have negative consequences. In this field, like in that of biology in general, nothing is absolute and all is relative. Generalizations should be avoided. When judging such and such a new foodstuff or such and such a new method, we must not be prisoners of 'a priori' ideas or routine techniques. The philosopher Alain said 'An idea is false as soon as one contents oneself with it'. Many scientists and legislators should meditate on these wise words.

Acknowledgment. Grateful acknowledgment is made to Mrs. Rerat for her translation of this paper.

References

1. Akiba Y, Matsumoto T (1975) Effet de la goitrine sur le bilan azoté du rat adulte. Eiyo To Shokuryo 28:457–461
2. Anonyme (1979) Régulateurs de la fécondité d'origine végétale. Chronique OMS 33:66–68
3. Bandyopadhyay GC (1970) The nature of the antivitamin E of Kidney beans fed to sheep. Dissert Abstr Internl 31:3090 B
4. Bender AE (1981) Low digestibility of legume nitrogen. Proceed Nut Soc Abstr communications March
5. Bender AE (1981) Toxicity of legumes and the assessment of protein quality. Proceed Nut Soc Abstr communications July
6. Bickoff EMA (1958) Plant oestrogens. J Agric Food Chem 10:410–413
7. Biggers JD (1959) Plant Phenols possessing oestrogenic Activity in: The Pharmacology of Plant Phenolics. Fairbairn (ed) Academic Press London, New York, pp 51–69
8. Bingefors S (1975) Antinutritional substances in peas. Sver Utsaedesfoeren Tidskr 85:211–219
9. Bishara HN, Walker HF (1977) Status of vitamin B_6 of pig with diet containing linseed meal. Br J Nutr 37:321–331
10. Bleumink E (1970) Food Allergy. The chemical nature of the substances eliciting symptoms. World Rev Nutr & Diet 12:505–570
11. Campbell LD, Smith IK (1979) Responses of growing chickens to high dietary content of rapeseed meal. Brit Poult Sci 28:231–237
12. Chaumeil JC, Carouge ML, Ardison J (1979) Etude de la teneur en vicine de diverses variétés de *Vicia faba*. Toxicol European Res 2:253–256
13. Dijkhof J, Poort SR, Poort C (1977) Effect of feeding soybean flour containing diets on the protein synthetic pattern of the rat pancreas. J Nutr 107:1985–1995
14. Dikkhof J, Poort C (1978) Changes in rat pancreatic synthesis after a single feeding with diets containing raw or heated soybeans. J Nutr 108:1222–1228
15. Drane HM, Patterson DSP, Roberts BA, Saba N (1980) Estrogenic activity of soybeans products. Food Cosmet Toxicol 18:425–427
16. Dupuy HP, Lee JG (1954) The isolation of a material capable of producing experimental Lathyrism. J Am Pharm Assoc 43:61–62
17. Ferrando R (1961) Alimentation et Equilibre Biologique. Collection de Philsophie Scientifique Flammarion edit Paris
18. Ferrando R, Guilleux Marie Madeleine, Guerillot-Vinet Andrée (1961) Teneur en oestrogènes des plantes en fonction des conditions de culture. Nature (London) 192:1205–1286
19. Ferrando R (1979) Aliments traditionnels et non traditionnels. Collection FAO: Alimentation et Nutrition No 2 Rome
20. Galloway DH (1973) Gas-Forming property of food legumes. In: Nutritional Improvement of Food Legumes by Breeding. PAG-ONU (eds) 263–270
21. Goel R, Verma J (1981) Identification of flatulence factors in some pulses. Curr Sci 50:144
22. Gontzea I, Ferrando R, Sutzesco P (1968) Substances antinutritives naturelles des aliments. Vigot (ed) Paris

23. Griffiths DW (1979) The inhibition of digestive enzymes by extracts of field bean – *Vicia faba*. J Sci Food Agric 30:458–462
24. Honovar PM, Shih CV, Liener IE (1962) Inhibition of growth of rats by purified hemagglutinin fraction isolated from *phaseolus vulgaris*. J Nutr 77:109–116
25. Jaffe WG (1969) Hemagglutinin. In: Liener IE (ed) Toxics Constituents of Plant Foodstuffs. Academic Press New York, pp 69–81
26. Kienholz EW (1962) Evidence for chick growth inhibitors in several legume seeds. Poultry Sci 41:367–371
27. Kilshaw PS, Sissons JW (1979) Gastrointestinal allergy to soybean protein in preruminant calves. Allergic constituents of soybean products. Res Vet Sci 27: 366–371
28. Klosterman HJ (1974) Vitamin B_6 Antagonists of Natural origin. J Agr Food Chem 22:13–16
29. Kunitz M (1945) Crystallization of a trypsin inhibitor from Soybean. Science. 101:668–670
30. Liener IE (1966) Toxic substances associated with seed protein. In: World Protein Resources Adv Chemistry Series No 57, pp 178–191
31. Liener IE (1973) Antitryptic and other antinutritional factors in legumes. In: Nutritional Improvement of Food Legumes by Breeding. PAG-ONU 239–258
32. Lindner HR (1976) Occurence of anabolic agents in plants and their importance. In: Anabolic Agents in Animal Production, Lu F and Rendel J (eds), Georg Thiem Publ, Stuttgart, pp 151–158
33. McGinnis J (1973) Investigations of Flatulence properties of different food products. In: Nutritional Improvement of Food Legumes by Breeding. PAG-ONU, pp 271–272
34. Moiseenak AG, Smashevskii ND, Androsyuk GV, Bud'ko TN, Grinevich VP (1975) Ant vitamin properties of pisamine (natural factor from sown peas) in relation to pantothenic acid in rats. Vesti Akad Navuk B URSS Ser Biyal Navuk 3:68–72
35. Osborne TB, Mendel LB (1912) Beobactungen Über Wachstum die Futteringsuchen mit isolierten Nahrungsubstanzen. Z Phyisol Chem 80:307–314
36. Pag (1973) Nutritional Improvement of Food Legumes by Breeding. Symposium at FAO Rome 1972. Published 1973 PAG-ONU
37. Papas A, Ingalls JR, Campbell D (1979) Studies on the effects of rapeseed meal on thyroid status of cattle, glucosinolate and iodine content of milk and other parameters. J Nutr 109:1129–1139
38. Piva G, Cardillo M, Rastelli A, Jacini G (1974) Changes in Brassica napus progoitrin induced by sheep rumen fluid. J Amer Oil Chem Soc 51:517–518
39. Rackis JJ (1981) Significance of soya trypsin inhibitors in nutrition. J Amer Oil Chem Soc 58:495–501
40. Richardson M (1981) Protein inhibitors of enzymes. Food Chem 6:235–253
41. Sapeika N (1969) Food Pharmacology. Thomas (ed) Springfield
42. Somogyi JC (1978) Natural toxic substances in food in toxicology & Nutrition. Truhaut R and Ferrando R (eds), World Rev Nutr Diet 29:42–59
43. Sotelo A, Arteaga ME, Frias MI, Gonzales-Garza MI (1980) Cytotoxic effects of two legumes in epithelial cells of the small Intestine. Qual Plant Plant Foods Hum Nutr 30:79–85
44. Steffen R, Fox FR, Kassel B (1978) Effect of Trypsin inhibitors on growth and metamorphosis of Corn Borer Larvae *Ostrinia nubilis* (Hübner). J Agric Food Chem 26:170–174
45. Turner RH, Liener IE (1975) Effect of selective elimination of Hemagglutinins on nutritive value of soya. J Agric Food Chem 23:484–487
46. Valdebouze P (1977) Activités antitrypsiques et hémaglutinantes dans les graines de quelques espèces de légumineuses. Comm Europ Communities EUR 5686:87–98
47. Vermorel M, Fayet JC, Baudet JJ (1978) Valorisation du Tourteau de Colza I – Influence de l'élimination des glucosinolates, du dépelliculage et de l'extraction des galactosides sur sa valeur nutritive. Etude sur le rat en croissance. Ann Biol Anim Bioch Biophys 18:1393–1412

Round Table Conclusions

ANDRÉ FRANCOIS

CNERNA, 72 rue de Sèvres 75007 Paris, France

Although in the course of this congress we have acquired an extensive amount of high level scientific and technical knowledge, we have also acquired some bits of knowledge. This was obligatory; each specialist excels in his sector and gave us the benefit of his experience but the time has now come to attempt a synthesis. In our Round Table Discussions, we have gone one step further since we are trying to utilize these bits of knowledge to solve concrete problems. We have dealt with problems which people concerned with economics have to tackle. In fact the theme proposed here is that of applications of plant proteins and their use in man's food.

It is not my aim to sum up what has been so well said in the various presentations; rather I shall endeavour to apply their know-how to the problem with which we are dealing and place those data in a more general frame.

Choosing food and the resulting purchases of food essentially obeys four parameters:

(1) physiological need (hunger);
(2) pleasure and quality of life (products must be tasty and easy to prepare);
(3) fear (fear of a bad effect of food on health, a negative factor);
(4) cost and availability of products.

Purchase of food always results from the association of these parameters. Whether one or the other plays a greater part, the purchase will be either festive or just plain. Moreover, choices of food are more or less guided by reason and knowledge (mainly dietetic) or personal tastes (chiefly gustative). In all cases, price plays an important part in the final decision.

In short, choosing food is subordinate to a series of examinations. Buyers' concerns are taken care of either by the buyers themselves, or by the Government. Appreciation of hygienic qualities as well as food-value is the first consideration. It is obvious that all products on sale must be innocuous. However, the data to be estimated are hidden. This is the first problem which dictated the rules based on scientific knowledge collected in the fields of toxicology, microbiology and virology. As for food-value, purchases depend on knowledge. Proper labelling, informs the consumer and helps him in his choice. Finally, as far as gustative qualities are concerned, buyers' own likings guide their purchases.

How shall we come to an appreciation of the qualities of plant proteins through these various criteria? As for hygienic qualities, it goes without saying that food must be free from any toxic element and that bacteriological quality must be guaranteed. Food-value depends on technological treatments undergone. For example heating can deteriorate the quality of proteins. In all cases, testing is a must. Regulations can take care of quality standards. One must underline here an important advantage of plant protein: the percentage of fats can be controlled. However, taste of raw products (raw, extracted, textured, spun proteins) is not very palatable. It is therefore necessary to have these products undergo very elaborate culinary preparations and if possible without reference to traditional mixtures based on meat or fish.

Whether or not these elaborate products are acceptable to the buyers is obviously a fundamental characteristic, to which the commercial success of a product is subordinate. From an economical point of view, one may wonder whether such a product presents any practical interest, whether it answers a need, and whether it can be marketed.

As far as transformed plant proteins are concerned, the answer is definitely yes. Such a product can provide the consumer with cheap protein. In world crises, the lowering of food costs can only be a reasonable aim. Moreover, plant proteins can contribute to a better world economy. As a matter of fact, meat production is costly when taking into account the low yield of the transformation of plant proteins into animal proteins. It should not be inferred, however, that transformed plant proteins may offer a solution to hunger throughout the world. They must be part of food in developed as well as developing countries and acceptable to both types of populations. In this respect, the security in food supplies contributed by plant proteins should be underlined. As a matter of fact, they can help facilitate a better world distribution of proteins. Their manufacture may be adjusted to the nature of the raw material available (e.g., soya bean, sunflower, rape, lentils, beans, olives, green leaves), the production of which depends on the climate and agronomical and economical conditions.

This Congress has shown that all problems are not solved. Yet I can measure the ground covered. Ten years ago in France, the Délégation Générale de la Recherche Scientifique et Technique initiated a large research on transformed plant proteins. I had the honour to instigate it with my colleague J. Mossé.

We were able to organize an important united action together with scientists chosen in the fields of technology, food, and economics. Two main industrial partners – Rhone Poulenc, les Moulins à Fèves de France – conducted a pilot project and even went as far as industrial realization.

On a general level, our aim was to bring the interest of these products to the awareness of the administration and the industrial sphere. The object was to utilize raw material produced on national soil since technology used in the case of soya bean was already known. Your presentations show the extraordinary progresses registered throughout the world, during the past decade, both in basic knowledge in physico-chemical properties of various proteins and in the technologies set forth.

Yet there are restraints which are by nature permanent. I mean that new technologies call for careful attention. For instance hygiene and food value of any new product must be carefully tested.

It is often thought that regulations hamper technological innovation. This is not necessarily true. When the presentation file of a new product is well made and contains all scientific data showing innocuousness and food-value, the authorization for marketing is given without delay.

Information plays a great part. Consumers must be ready to understand technical information. Formation of young citizens must start early. This is what we observed in France: nutrition can be taught in elementary schools.

However, some aspects call for immediate solutions. The addition of plant proteins to culinary preparations served in cafeterias calls for reservations because of no labelling. In this case consumers must be informed. It must be rightly emphasized that confidence between producer and consumer is a necessary condition for the commercial success of plant proteins. It goes without saying that the existence of satisfactory testing methods must assist or raise this confidence.

In short, industrial plant proteins cannot be considered as a cure-all. They are, however, an interesting product both for their food value and the part they can play to improve our food economy.

In the daily consumption of food, one exercises one's responsibility and freedom. Every buyer must be able to choose according to the criteria which I recalled at the beginning of this discussion. The type of products to which we have referred is not an exception. Hygiene, food value, practicability, low cost are all incentives to buy. None of them must be neglected.

Permit me to thank and congratulate all those who contributed in this conference to a greater awareness of plant proteins usable in man's food.